T0234789

SpringerBriefs in Applied Sciences and Technology

Computational Intelligence

Series Editor

Janusz Kacprzyk, Systems Research Institute, Polish Academy of Sciences, Warsaw, Poland

SpringerBriefs in Computational Intelligence are a series of slim high-quality publications encompassing the entire spectrum of Computational Intelligence. Featuring compact volumes of 50 to 125 pages (approximately 20,000-45,000 words), Briefs are shorter than a conventional book but longer than a journal article. Thus Briefs serve as timely, concise tools for students, researchers, and professionals.

More information about this subseries at http://www.springer.com/series/10618

Oscar Castillo · Patricia Ochoa · Jose Soria

Differential Evolution Algorithm with Type-2 Fuzzy Logic for Dynamic Parameter Adaptation with Application to Intelligent Control

 Springer

Oscar Castillo
Division of Graduate Studies
Tijuana Institute of Technology
Tijuana, Mexico

Patricia Ochoa
Division of Graduate Studies
Tijuana Institute of Technology
Tijuana, Mexico

Jose Soria
Division of Graduate Studies
Tijuana Institute of Technology
Tijuana, Mexico

ISSN 2191-530X   ISSN 2191-5318   (electronic)
SpringerBriefs in Applied Sciences and Technology
ISSN 2625-3704   ISSN 2625-3712   (electronic)
SpringerBriefs in Computational Intelligence
ISBN 978-3-030-62132-2   ISBN 978-3-030-62133-9   (eBook)
https://doi.org/10.1007/978-3-030-62133-9

This Springer imprint is published by the registered company Springer Nature Switzerland AG
The registered company address is: Gewerbestrasse 11, 6330 Cham, Switzerland

Preface

This book was mainly based on the combining fuzzy logic and the differential evolution algorithm, which are used to obtain better results in controllers of non-linear plants. In this book, we test the proposed method using five benchmark control problems. First, the water tank, temperature, mobile robot, and inverted pendulum controllers are presented. For these four controllers, experimentation is carried out using a type-1 fuzzy system and an interval type-2 fuzzy system. The last controller is the DC motor, and in this case, the experiment was performed with type-1, interval type-2, and generalized type-2 fuzzy systems. When we use the fuzzy system combined with the algorithm, we can note that the results obtained in each of the used controllers are better, and as the uncertainty increases, the results are even better. For this reason, we consider that the proposed method using fuzzy systems, fuzzy controllers, and differential evolution algorithm improves the behavior of the complex control problems.

This book is intended to be a reference for scientists and engineers interested in applying fuzzy logic techniques for solving problems in intelligent control. This book can also be used as a reference for graduate courses like the following: soft computing, swarm intelligence, bio-inspired algorithms, intelligent control, fuzzy control, and similar ones. We consider that this book can also be used to get novel ideas for new lines of research, or to continue the lines of research proposed by the authors of the book.

In Chap. 1, we begin by offering a brief introduction of the potential use of the optimization strategies in different real-world applications. The chapter contains information relevant to the different concepts used in the book as well as important articles of the researchers working with fuzzy logic, bio-inspired algorithms, or both, as well as references of the differential evolution algorithm, which have served us as guide to make this book

We describe in Chap. 2 the basic concepts, notation, and theory of fuzzy logic and fuzzy controllers. This chapter overviews the background, main definitions, and basic concepts, useful for the development of this research work.

Chapter 3 contains the description of the differential evolution algorithm, the equations that involve the development of the algorithm, as well as a detailed explanation of each step of the algorithm.

We present in Chap. 4 the structure of the proposed methodology, the combination of the differential evolution algorithm and fuzzy logic, and the optimization of the parameters that are dynamic during the execution of the algorithm is described.

Chapter 5 presents the different control problems used for experimentation, a description is made of each control problem, the fuzzy system it contains, and the rules. In addition, contains the objective function, the mathematical equations of the used fuzzy systems, the results obtained for each of the controllers, the comparison of the different variants used, as well as the statistical tests performed for the comparison of the original algorithm and the different variants using type-1, interval type-2 fuzzy, and generalized type-2 fuzzy systems.

We describe in Chap. 6 the conclusions of this work, it is explained which were the controllers that presented greater or lesser difficulty for the proposed methodology, conclusions are given based on the tests performed with the original algorithm and the different variants as well as those performed with other meta-heuristics which also use the same concept of making certain parameters dynamic using fuzzy logic. Our conclusions are divided into two main areas, the first corresponding to all the experimentation carried out in four control problems using a fuzzy system of one input and one output which are type-1 and interval type-2 fuzzy systems. On the other hand, the experimentation carried out with a control problem with a fuzzy system of two inputs and two outputs which are type-1, interval type-2 fuzzy, and generalized type-2 fuzzy systems.

We end this preface of the book by giving thanks to all the people who have help or encourage us during the writing of this book. First of all, we would like to thank Dr. Patricia Melin, for always supporting our work and for motivating us to write our research work. We would also like to thank our colleagues working in soft computing, which are too many to mention each by their name. Of course, we need to thank our supporting agencies, CONACYT and TNM, in our country for their help during this project. We have to thank our institution, Tijuana Institute of Technology, for always supporting our projects. Finally, we thank our respective families for their continuous support during the time that we spend in this project.

Tijuana, Mexico   Prof. Oscar Castillo
                                                                              Dr. Patricia Ochoa
                                                                              Prof. Jose Soria

Contents

Chapter 1
Introduction

The use of algorithms based on nature has become very common in evolutionary computation and metaheuristics. In this book we propose to use one of these algorithms, in particular the Differential Evolution (DE) integrating fuzzy logic to dynamically adapt some of its parameters during execution.

The main idea of fuzzy set theory was originally proposed by Zadeh in 1965 and was first applied to control theory in 1974 by Mamdani [1–5]. Based on these works, fuzzy controllers have been successfully applied in numerous applications [6–9]. In [10], an analytical structure and stability analysis of a fuzzy PID controller is presented. Recently, the DE algorithm has been applied in different problems like: [11–19] just to mention some recent works.

Similarly, the use of fuzzy logic for different applications has been increasingly relevant, and we can find the use of fuzzy logic in areas such as medicine, control, robotics, artificial intelligence, being the control area our main interest [20–23]. In our case, we focus for this paper on the combination of fuzzy logic and metaheuristic algorithms, since it has been demonstrated that both methodologies when combined improve the performance of the algorithms, but in particular the use of Type-2 fuzzy system can improve even more the performance, and to mention some related works of interest [24–34].

The design of fuzzy controllers is an important application area for metaheuristic algorithms. For this book we use the Differential Evolution algorithm enhanced with Type-1, Interval Type-2 fuzzy and Generalized Type-2 fuzzy systems for dynamic parameter adjustment and its application to fuzzy system design for control problems. A comparison is made among the original algorithm, the proposed algorithm and finally noise was also added to the control system to test the proposed algorithm.

The nonlinear characteristics of ill-defined and complex modern plants make classical controllers inadequate for such systems because they require complicated mathematical models. However, the use of Fuzzy sets and fuzzy logic principles have enabled researchers to understand better and hence control, complex systems that are

1

O. Castillo et al., *Differential Evolution Algorithm with Type-2 Fuzzy Logic for Dynamic Parameter Adaptation with Application to Intelligent Control*, SpringerBriefs in Computational Intelligence, https://doi.org/10.1007/978-3-030-62133-9_1

difficult to model. These newly developed fuzzy logic controllers have given control systems a certain degree of intelligence and flexibility.

References

1. Mamdani EH (1974) Application of fuzzy algorithms for control of simple dynamic plant. In Proc Inst Electr Eng 121(12):1585
2. Zadeh LA (1975) The concept of a linguistic variable and its application to approximate reasoning—I. Inf Sci 8(3):199–249
3. Zadeh L (1978) Fuzzy sets as a basis for a theory of possibility. Fuzzy Sets Syst 1(1):3–28
4. Zadeh LA (1975) The concept of a linguistic variable and its application to approximate reasoning—II. Inf Sci 8(4):301–357
5. Dubois D, Sandri SA (1997) Editorial. Fuzzy Sets Syst 90(2):109–110
6. Gao Q (2017) Universal fuzzy models and universal fuzzy controllers for stochastic non-affine nonlinear systems. In: Universal fuzzy controllers for non-affine nonlinear systems. Springer Singapore, Singapore, pp 45–70
7. Radu-Emil P, Radu-Codrut D, Emil MP (2017) Grey wolf optimizer algorithm-based tuning of fuzzy control systems with reduced parametric sensitivity. IEEE Trans Ind Electron 64(1):527–534
8. Van-Binh B, Quy-Cao T, Hai-Le B (2017) Multi-objective optimal design of fuzzy controller 46 for structural vibration control using Hedge-algebras approach. Artif Intell Rev 43:345–379
9. Driankov D, Palm R (2013) Advances in fuzzy control. Physica
10. Mohan BM, Sinha A (2008) Analytical structure and stability analysis of a fuzzy PID controller. Appl Soft Comput 8(1):749–758
11. Chia-Feng J, Ying-Han C, Yue-Hua J (2015) Wall-following control of a hexapod robot using a data-driven fuzzy controller learned through differential evolution. IEEE Trans Ind Electron 62(1):611–619
12. Bi Y, Srinivasan D, Lu X, Sun Z, Zeng W (2014) Type-2 fuzzy multi-intersection traffic signal control with differential evolution optimization. Expert Syst Appl 41(16):7338–7349
13. Sun Z, Wang N, Srinivasan D, Bi Y (2014) Optimal tunning of type-2 fuzzy logic power system stabilizer based on differential evolution algorithm. Int J Electr Power Energy Syst 62:19–28
14. Liu J, Lampinen J (2005) A fuzzy adaptive differential evolution algorithm. Soft Comput 9(6):448–462
15. Tang L, Zhao Y, Liu J (2014) An improved differential evolution algorithm for practical dynamic scheduling in steelmaking-continuous casting production. IEEE Trans Evol Comput 18(2):209–225
16. Wang Y, Liu Z-Z, Li J, Li H-X, Wang J (2018) On the selection of solutions for mutation in differential evolution. Front Comput Sci 12(2):297–315
17. Salehpour M, Jamali A, Bagheri A, Nariman-zadeh N (2017) A new adaptive differential evolution optimization algorithm based on fuzzy inference system. Eng Sci Technol Int J 20(2):587–597
18. Aalto J, Lampinen J (2014) A mutation and crossover adaptation mechanism for differential evolution algorithm. In: 2014 IEEE Congress on evolutionary computation (CEC), Beijing, China, pp 451–458
19. Aalto J, Lampinen J (2013) A mutation adaptation mechanism for Differential Evolution algorithm. In: 2013 IEEE Congress on evolutionary computation, Cancun, Mexico, pp 55–62
20. Sa-ngiamvibool W (2017) Optimal fuzzy logic proportional integral derivative controller design by bee algorithm for hydro-thermal system. IEEE Trans Ind Inform 1–1
21. Cuevas E, Luque A, Zaldívar D, Pérez-Cisneros M (2017) Evolutionary calibration of fractional fuzzy controllers. Appl Intell 47, pp 291–303

22. Caraveo C, Valdez F, Castillo O (2016) Optimization of fuzzy controller design using a new bee colony algorithm with fuzzy dynamic parameter adaptation. Appl Soft Comput 43:131–142

23. Noshadi A, Shi J, Lee WS, Shi P, Kalam A (2016) Optimal PID-type fuzzy logic controller for a multi-input multi-output active magnetic bearing system. Neural Comput Appl 27(7):2031–2046

24. Martínez-Soto R, Castillo O, Castro JR (2014) Genetic algorithm optimization for type-2 non-singleton fuzzy logic controllers. In: Recent advances on hybrid approaches for designing intelligent systems, vol 547. Springer, Germany

25. Melin P, Astudillo L, Castillo O, Valdez F, Garcia M (2013) Optimal design of type-2 and type-1 fuzzy tracking controllers for autonomous mobile robots under perturbed torques using a new chemical optimization paradigm. Expert Syst Appl 40(8):3185–3195

26. De la O D, Castillo O, Soria J (2007) Optimization of reactive control for mobile robots based on the CRA using type-2 fuzzy logic. In: Melin P, Castillo O, Kacprzyk J (eds) Nature-inspired design of hybrid intelligent systems, vol 667. Springer International Publishing, Cham, pp. 505–515

27. Olivas F, Valdez F, Melin P, Sombra A, Castillo O (2019) Interval type-2 fuzzy logic for dynamic parameter adaptation in a modified gravitational search algorithm. Inf Sci 476:159–175

28. Castillo O, Melin P, Valdez F, Soria J, Ontiveros-Robles E, Peraza C, Ochoa P (2019) Shadowed type-2 fuzzy systems for dynamic parameter adaptation in harmony search and differential evolution algorithms. Algorithms 12(1):17

29. Ontiveros-Robles E, Melin P, Castillo O (2018) Comparative analysis of noise robustness of type 2 fuzzy logic controllers. Kybernetika 54(1):175–201

30. Castillo O, Cervantes L, Soria J, Sanchez M, Castro JR (2016) A generalized type-2 fuzzy granular approach with applications to aerospace. Inf Sci 354:165–177

31. Castillo O, Amador-Angulo L, Castro JR, Garcia-Valdez M (2016) A comparative study of type-1 fuzzy logic systems, interval type-2 fuzzy logic systems and generalized type-2 fuzzy logic systems in control problems. Inf Sci 354:257–274

32. Sánchez MA, Castillo O, Castro JR (2015) Information granule formation via the concept of uncertainty-based information with interval type-2 fuzzy sets representation and Takagi–Sugeno–Kang consequents optimized with Cuckoo search. Appl Soft Comput 27:602–609

33. Cervantes L, Castillo O (2015) Type-2 fuzzy logic aggregation of multiple fuzzy controllers for airplane flight control. Information Sci 324:247–256

34. Amador-Angulo L, Castillo O (2015, June) Statistical analysis of type-1 and interval type-2 fuzzy logic in dynamic parameter adaptation of the BCO. In: 2015 conference of the International Fuzzy Systems Association and the European Society for Fuzzy Logic and Technology (IFSA-EUSFLAT-15). Mendel JM, Liu X (2013 December) Simplified interval type-2 fuzzy logic systems. IEEE Trans Fuzzy Syst 21(6):1056–1069

Chapter 2
Fuzzy Logic Systems

In this chapter we present some basic concepts about fuzzy systems needed to better understand the ideas and the concepts of this book.

A fuzzy logic system (FLS) that is defined entirely in terms of Type-1 fuzzy sets is known as a Type 1 fuzzy logic system (Type-1 FLS) [1], and its elements are defined in Fig. 2.1.

A Type-2 fuzzy set, $\tilde{A}$, is characterized by Eq. 2.1:

$$\tilde{A} = \left\{ (x, u), u_{\tilde{A}}(x, u) | \forall x \in X, \forall u \in J_x \subseteq [0, 1] \right\} \tag{2.1}$$

where $0 \leq u_{\tilde{A}}(x, u) \leq 1$.

In a general way we can say that a type-2 fuzzy set is a generalization of a fuzzy set that is associated with a secondary source of uncertainty related to the definition of a set $\tilde{A}$. This additional source of uncertainty is represented as a secondary membership function. A Type-2 fuzzy system, whose secondary membership function is an interval, hence the name of Interval Type-2 fuzzy system, is expressed by two membership functions, where one represents the degree of belonging in X and the other gives a weighting to each of the Type-1 fuzzy systems. Figure 2.2 shows the architecture of an Interval Type-2 fuzzy system.

The output processor includes a type-reducer and a defuzzifier; it generates a Type-1 fuzzy set output (from the type-reducer) or a crisp number (from the defuzzifier).

The main difference between the Type-1 and the Interval Type-2 fuzzy systems is the footprint of uncertainty (FOU) that defines the uncertainty of $\tilde{A}$ as the union of all the primary belongings which is limited by two membership functions: an upper membership function and lower membership function. In addition, in the Interval Type-2 fuzzy system there is the defuzzifier block of a Type-1 is replaced by the processing block output consisting of a type-reducer followed by defuzzifier. It is important to mention that the inference operation in the Interval Type-2 fuzzy system is much more complicated than in Type-1 fuzzy system and that a type reducer is necessary for a Type-2 fuzzy system to convert Type-2 fuzzy sets into Type-1. These

© The Author(s), under exclusive license to Springer Nature Switzerland AG 2021
O. Castillo et al., *Differential Evolution Algorithm with Type-2 Fuzzy Logic for Dynamic Parameter Adaptation with Application to Intelligent Control*,
SpringerBriefs in Computational Intelligence,
https://doi.org/10.1007/978-3-030-62133-9_2

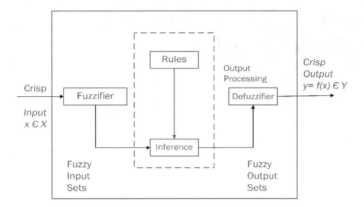

Fig. 2.1 Type-1 fuzzy logic system

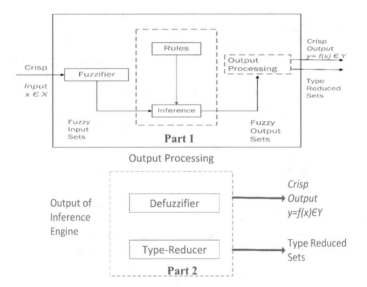

Fig. 2.2 Type-2 fuzzy logic system

differences are the ones that make the calculations of an Interval Type-2 fuzzy system more challenging.

With Generalized Type-2 fuzzy logic systems the logic is generally the same as for Type-1 and the Interval Type-2 fuzzy systems, but their operations are somewhat different, due to the nature of Generalized Type-2 fuzzy logic [1]. There are several mathematical definitions of a Generalized Type-2 fuzzy logic system, and we used the representation based on [2, 3] to define Generalized Type-2 Fuzzy Sets and are defined by the following Eq. 2.2:

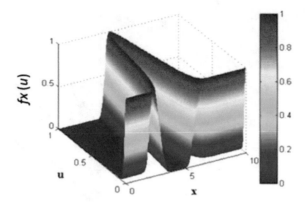

Fig. 2.3 Generalized type-2 membership function

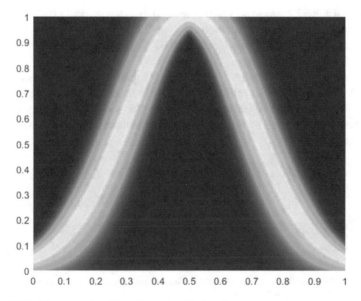

Fig. 2.4 FOU of the generalized type-2 membership function

$$\tilde{\tilde{A}} = \left\{ \left((x, u), \mu_{\tilde{A}}(x, u) \right) | \forall_x \in X, \quad \forall_u \in J_x \subseteq [0, 1] \right\} \tag{2.2}$$

where $J_x \subseteq [0, 1]$, x is the partition of the primary membership function, and u is the partition of the secondary membership function. In Fig. 2.3 we can find a representation of a Generalized Type-2 membership function, and in Fig. 2.4, the footprint of uncertainty (FOU) is illustrated, which is associated with the third dimension and allows a better modeling of real world uncertainty. It must be noted that there is a small difference in notation when compared with Type-1 and Interval Type-2 fuzzy system, this is, Type-1 and Interval Type-2 fuzzy system use the notation $\mu(x)$, but

Fig. 2.5 An example of the
associated type-2 fuzzy set
for the alpha-plane

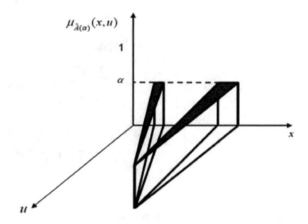

Generalized Type-2 fuzzy logic systems uses $f_x(u)$, in the vertical axis, and this is due
to the complexity involved in Generalized Type-2 fuzzy logic systems in comparison
with the others, as well as how Generalized Type-2 fuzzy logic systems has been
described in the literature.

The α-plane for a Generalized Type-2 fuzzy logic, in this case $\widetilde{A}$, is denoted by
$\widetilde{A}_\alpha$, and it is the union of all primary membership functions of $\tilde{A}$, which secondary
membership degrees are higher or equal to α ($0 \leq \alpha \leq 1$) [4, 5]. The equation of an
alpha plane is represented by Eq. 2.3. In Fig. 2.5 the representation of an alpha plane
is illustrated.

$$\widetilde{A}_\alpha = \left\{ (x, u), \mu_{\tilde{A}}(x, u) \geq \alpha | \forall_x \in X, \forall_u \in J_x \subseteq [0, 1] \right\} \tag{2.3}$$

References

1. Sánchez MA, Castillo O, Castro JR (2015) Generalized type-2 fuzzy systems for controlling
   a mobile robot and a performance comparison with interval type-2 and type-1 fuzzy systems.
   Expert Syst Appl 42(14):5904–5914
2. Melin P, González CI, Castro JR, Mendoza O, Castillo O (2014) Edge-detection method for
   image processing based on generalized type-2 fuzzy logic. IEEE Trans Fuzzy Syst 22(6):1515–
   1525
3. Sanchez MA, Castro JR, Castillo O (2013, April) Formation of general type-2 Gaussian membership functions based on the information granule numerical evidence. In: 2013 IEEE workshop
   on hybrid intelligent models and applications (HIMA), pp 1–6. IEEE
4. Castillo O, Atanassov K (2019) Comments on fuzzy sets, interval type-2 fuzzy sets, general
   type-2 fuzzy sets and intuitionistic fuzzy sets. I: Recent advances in intuitionistic fuzzy logic
   systems, pp 35–43. Springer, Cham
5. Mendel JM, Liu F, Zhai D (2009) Alpha-plane representation for type-2 fuzzy sets: theory and
   applications. IEEE Trans Fuzzy Syst 17(5):1189–1207

Chapter 3
Differential Evolution Algorithm

In this chapter, the description of the Differential Evolution algorithm is explained.

Differential Evolution is basically composed of 4 steps [1]: initialization, mutation, crossing and selection.

This is a non-deterministic technique based on the evolution of a vector population (individuals) of real values representing the solutions in the search space. The generation of new individuals is carried out by the differential crossover and mutation operators. The operation of the algorithm is explained in more detail below.

3.1 Population Structure

The Differential Evolution algorithm maintains a pair of vector populations, both of which contain Np D-dimensional vectors of real-valued parameters [2], the current population, symbolized P_x, is composed of the vector $x_{i,g}$ in Eqs. 3.1 and 3.2.

$$P_{x,g} = (x_{i,g}), i = 0, 1, \ldots, Np - 1, g = 0, 1, \ldots, g_{max},$$
$$x_{i,g} = (x_{j,i,g}), \quad j = 0, 1, \ldots, D - 1 \tag{3.1}$$

$$P_{v,g} = (v_{i,g}), i = 0, 1, \ldots, Np - 1, g = 0, 1, \ldots, g_{max},$$
$$v_{i,g} = (v_{j,i,g}), \quad j = 0, 1, \ldots, D - 1 \tag{3.2}$$

Each vector in the current population is recombined with a mutant vector to produce a trial population, P_u, the NP, Eq. 3.3 represents the mutant vector $u_{i,g}$:

$$P_{u,g} = (u_{i,g}), i = 0, 1, \ldots, Np - 1, g = 0, 1, \ldots, g_{max},$$
$$u_{i,g} = (u_{j,i,g}), \quad j = 0, 1, \ldots, D - 1 \tag{3.3}$$

© The Author(s), under exclusive license to Springer Nature Switzerland AG 2021
O. Castillo et al., *Differential Evolution Algorithm with Type-2 Fuzzy Logic for Dynamic Parameter Adaptation with Application to Intelligent Control*,
SpringerBriefs in Computational Intelligence,
https://doi.org/10.1007/978-3-030-62133-9_3

3.2 Initialization

Before initializing the population in Eq. 3.4, the upper and lower limits for each parameter must be specified. These 2D values can be collected by two initialized vectors, D-dimensional, b_L y b_U, for which the subscripts L and U indicate the lower and upper limits respectively. Once the initialization limits have been specified a number generator randomly assigns each parameter in every vector a value within the set range. For example, the initial value (g = 0) of the j-th vector parameter is ith:

$$x_{j,i,0} = rand_j(0, 1) \cdot \left(b_{j,U} - b_{j,L}\right) + b_{j,L} \tag{3.4}$$

3.3 Mutation

In particular, the differential mutation uses a random sample Eq. 3.5 showing how to combine three different vectors chosen randomly to create a mutant vector.

$$v_{i,g} = x_{r_0,g} + F \cdot \left(x_{r_1,g} - x_{r_2,g}\right) \tag{3.5}$$

The scale factor, $F \in (0,1)$ is a positive real number that controls the rate at which the population evolves. While there is no upper limit on F, the values are rarely greater than 1.0.

3.4 Crossover

To complement the differential mutation search strategy, DE also uses uniform crossover. This is sometimes known as discrete recombination (dual). In particular, DE crosses each vector with a mutant vector expressed in Eq. 3.6:

$$u_{i,g} = u_{j,i,g} \begin{cases} v_{j,i,g} if \left(rand_j(0, 1) \leq CR \, or \, j = j_{rand}\right) \\ x_{j,i,g} \quad otherwise \end{cases} \tag{3.6}$$

3.5 Selection

If the test vector, $u_{i,g}$ has a value of the objective function equal to or less than, its target vector, $X_{i,g}$, it replaces the target vector in the next generation; otherwise, the target retains its place in population for at least another generation, represented in the Eq. 3.7.

$$x_{i,g+1} = \begin{cases} u_{i,g} \; if \; f\left(u_{i,g}\right) \leq f\left(x_{i,g}\right) \\ \quad x_{i,g} \; otherwise \end{cases} \tag{3.7}$$

The process of mutation, recombination and selection are repeated until the optimum is found, or a terminating pre criteria specified is satisfied. DE is a simple, but powerful search engine that simulates natural evolution, combined with a mechanism to generate multiple search directions based on the distribution of solutions in the current population. Each vector i in the population at generation G, $xi \in G$, called at this moment of reproduction as the target vector will be able to generate one offspring, called trial vector (u_i, G). This trial vector is generated as follows: First of all, a search direction is defined by calculating the difference between a pair of vectors r_1 and r_2, called "*differential vectors*", both of them chosen at random from the population. This difference vector is also scaled by using a user defined parameter called "$F \geq 0$". This scaled difference vector is then added to a third vector r_3, called "*base vector*". As a result, a new vector is obtained, known as the mutation vector. After that, this mutation vector is recombined with the target vector (also called parent vector) by using discrete recombination (usually binomial crossover) controlled by a crossover parameter $0 \leq CR \leq 1$ whose value determines how similar the trial vector will be with respect to the target vector. There are several DE variants. However, the most known and used is DE/rand/1/bin, where the base vector is chosen at random, there is only a pair of differential vectors and a binomial crossover is used.

Figure 3.1 corresponds to the pseudocode of the original Differential Evolution algorithm, which summarizes more clearly the equations mentioned above.

In summary, we can say that the differential Evolution algorithm involves defining a population of NP vectors, these vectors are initialized, the lower and upper limits have to be previously defined depending on the problem, 3 individuals are selected, mutation and a crossover operations are applied, If the value resulting is better than the one chosen for replacement, then it replaces it. Otherwise, the main individual is retained.

Generate the initial population of individuals
Do
    For each individual j in the population
    Choose three numbers x_1, x_2 and x_3 that is, $1 \leq x_0, x_1, x_2 \leq N$ with $x_0 \neq x_1 \neq x_2 \neq j$
    Generate a random integer $i_{rand} \in (1, N)$
      For each parameter i

$$v_{i,g} = x_{r_0,g} + F \cdot \left(x_{r_1,g} - x_{r_2,g} \right)$$

$$u_{i,g} = u_{j,i,g} \begin{cases} v_{j,i,g} \ if \ \left(rand_j(0,1) \leq Cr \ or \ j = j_{rand} \right) \\ \quad x_{j,i,g} \quad otherwise \end{cases}$$

      End For
    Replace $x_{j,i,g}$ with the child $u_{i,g}$ if $u_{i,g}$ is better
    End For
Until the termination condition is achieved

Fig. 3.1 Pseudocode of the differential evolution algorithm

References

1. Price K, Storn RM, Lampinen JA (2006) Differential evolution: a practical approach to global optimization. Springer Science & Business Media
2. Salehpour M, Jamali A, Bagheri A, Nariman-Zadeh N (2017) A new adaptive differential evolution optimization algorithm based on fuzzy inference system. Eng Sci Technol Int J 20(2):587–597

Chapter 4
Proposed Method

This chapter explains in a general way how the Differential Evolution algorithm combined with fuzzy logic is used, just as the structure shown of the proposed methodology.

For the experiments we use the proposed method, which we call DE+FS (Fuzzy Differential Evolution), which has dynamic adjustment of the parameters as main difference with respect to the original algorithm, which is recommended to be used in a range of [0, 1]. The proposed method makes the F (mutation) and CR (crossover) parameters to be moved dynamically during the execution of the algorithm within the range [0, 1]. We have previous works where we used our proposed DE+T1FS method for Benchmark problems and we obtained good results [1–4].

For this work we use our DE+FS method to change the parameter values of the membership functions for benchmark controller problems. A comparison is made where we dynamically move the F (mutation) or CR (crossover) parameters, with Type-1 fuzzy logic, Interval Type-2 fuzzy system or Generalized Type-2 fuzzy logic system, which is in charge of dynamically moving the F or CR parameters.

Figure 4.1 represents the general proposed methodology where the F and CR parameters represent mutation and crossover respectively, with the help of each of the different fuzzy systems they are dynamically changing during the execution of the algorithm. It is important to mention that for the experimentation, each one of the proposed fuzzy system was worked separately.

Equation 4.1 represents the current generations and is defined by the number of generation elapsed and maximum number of generations is defined by the number of generations established for DE to find the best solution, the outputs are F or CR parameters, are calculated in Eqs. 4.2 and 4.3:

$$Generations = \frac{Current\ Generation}{Maximun\ of\ generations} \tag{4.1}$$

© The Author(s), under exclusive license to Springer Nature Switzerland AG 2021
O. Castillo et al., *Differential Evolution Algorithm with Type-2 Fuzzy Logic for Dynamic Parameter Adaptation with Application to Intelligent Control*,
SpringerBriefs in Computational Intelligence,
https://doi.org/10.1007/978-3-030-62133-9_4

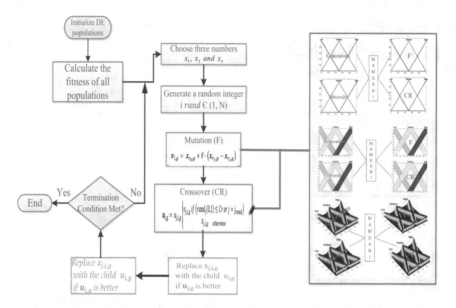

Fig. 4.1 Original algorithm with a fuzzy system

$$F = \frac{\sum_{i=1}^{r_F} \mu_i^F (F_{1i})}{\sum_{i=1}^{r_F} \mu_i^F} \tag{4.2}$$

$$CR = \frac{\sum_{i=1}^{r_{CR}} \mu_i^{CR} (CR_{1i})}{\sum_{i=1}^{r_{CR}} \mu_i^{CR}} \tag{4.3}$$

where mutation is F; r_F is the number of rules of the fuzzy system corresponding to F; F_{1i}, is the output result for rule i corresponding to F; μ_i^F, is the membership function of rule i corresponding to F, and crossover is CR; r_{CR}, is the number of rules of the fuzzy system corresponding to CR; CR_{1i}, is the output result for rule i corresponding to CR; μ_i^F, is the membership function of rule i corresponding to CR.

Figure 4.2 represents the pseudocode in which the fuzzy system is used to dynamically adapt F or CR parameters.

Generate the initial population of individuals
Do
    For *each individual j in the population*
    Choose three numbers x_1, x_2 and x_3 that is, $1 \leq x_0, x_1, x_2 \leq N$ with $x_0 \neq x_1 \neq x_2 \neq j$
    Generate a random integer $i_{rand} \in (1, N)$
        For *each parameter i*
        Calculate generation using Equations (4.1)
        Use a fuzzy system to calculate the new Mutation (4.2) or Crossover (4.3) parameters

$$v_{i,g} = x_{r_0,g} + F \cdot (x_{r_1,g} - x_{r_2,g})$$

$$u_{i,g} = u_{j,i,g} \begin{cases} v_{j,i,g} \ if \ (rand_j(0,1) \leq CR \ or \ j = j_{rand}) \\ x_{j,i,g} \quad otherwise \end{cases}$$

        End For
        Replace $x_{j,i,g}$ with the child $u_{i,g}$ if $u_{i,g}$ is better
    End For
Until the termination condition is achieved

Fig. 4.2 Pseudocode for the fuzzy differential evolution

References

1. Ochoa P, Castillo O, Soria J (2016, October) Type-2 fuzzy logic dynamic parameter adaptation in a new fuzzy differential evolution method. In: 2016 annual conference of the North American Fuzzy Information Processing Society (NAFIPS), pp 1–6. IEEE
2. Ochoa P, Castillo O, Soria J (2016, September) Fuzzy differential evolution method with dynamic parameter adaptation using type-2 fuzzy logic. In: 2016 IEEE 8th international conference on intelligent systems (IS), pp 113–118. IEEE
3. Ochoa P, Castillo O, Soria J (2017) Differential evolution using fuzzy logic and a comparative study with other metaheuristics. In: Melin P, Castillo O, Kacprzyk J (eds) Nature-inspired design of hybrid intelligent systems, vol 667. Springer International Publishing, Cham, pp 257–268
4. Castillo O, Ochoa P, Soria J (2016) Differential evolution with fuzzy logic for dynamic adaptation of parameters in mathematical function optimization. In: Angelov P, Sotirov S (eds) Imprecision and uncertainty in information representation and processing, vol 332. Springer International Publishing, Cham, pp 361–374

Chapter 5
Case Studies

This chapter proposes the use of the Differential Evolution algorithm with fuzzy logic for parameter adaptation in the optimal design of fuzzy controllers for non-linear plants. The Differential Evolution algorithm is enhanced using Type-1 and Interval Type-2 fuzzy systems for achieving dynamic adaptation of the mutation parameter. In this chapter four control optimization problems in which the Differential Evolution algorithm optimizes the membership functions of the fuzzy controllers are presented. First, the experiments were performed with the original algorithm, second the experiments were performed with the Fuzzy Differential Evolution (in this case the mutation parameter is dynamic) and the last experiments were performed applying noise to the control plant by using Fuzzy Differential Evolution were performed. The characteristics of each of the control problems to be used for experimentation are listed below.

5.1 Water Tank Controller

The first case study is of the water tank controller, whose main objective is controlling the water level in a tank, and Fig. 5.1 graphically represents the way in which the valve operates and hence the filling process in the tank [1, 2].

The mathematical model of the water tank controller is presented in the Eq. 5.1:

$$\frac{d}{dt}Vol = A\frac{dH}{dt} = bV - a\sqrt{H} \tag{5.1}$$

where:

Vol Volume of water in the tank
A   Cross-sectional area
b   Constant related to the flow rate into tank.

© The Author(s), under exclusive license to Springer Nature Switzerland AG 2021   17
O. Castillo et al., *Differential Evolution Algorithm with Type-2 Fuzzy Logic for Dynamic Parameter Adaptation with Application to Intelligent Control*,
SpringerBriefs in Computational Intelligence,
https://doi.org/10.1007/978-3-030-62133-9_5

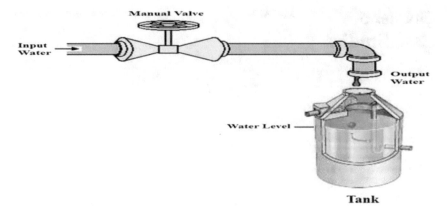

Fig. 5.1 Graphical illustration of the water tank controller

a   Constant related to the flow rate out of the tank.
H   Height of water H as a function of time

   Where *Vol* is the volume of water in the tank, *A* is the cross-sectional area of the tank, **b** is a constant related to the flow rate into the tank, and *a* is a constant related to the flow rate out of the tank. The equation describes the height of water *H* as a function of time, due to the difference between flow rates into and out of the tank.

   Figure 5.2 illustrates the structure of the fuzzy system for this control problem, the way in which the membership functions are granulated, the type of functions and how many membership functions correspond to each input and output. The 5 fuzzy rules of the controller are presented in Fig. 5.3.

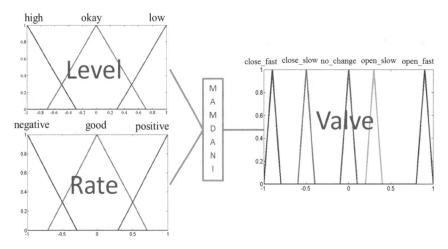

Fig. 5.2 Fuzzy control system for the water tank

1. If (level is okay) then (valve is no_change)
2. If (level is low) then (valve is open_fast)
3. If (level is high) then (valve is close_fast)
4. If (level is okay) and (rate is positive) then (valve is close_slow)
5. If (level is okay) and (rate is negative) then (valve is open_slow)

Fig. 5.3 Rules of the water tank controller

The fuzzy system designed for the water tank controller is built based on the actual filling behavior of a water tank, and the set of rules are made in terms of the theory of how this process is performed.

5.2   Temperature Controller

The second case study is a temperature controller, whose main objective is to control the temperature in the water flow. Figures 5.4 and 5.5 show the fuzzy system for the temperature controller, the system contains two inputs and two outputs, is of Mamdani type and the rules are presented in Fig. 5.6 [3].

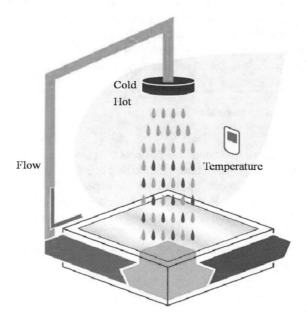

Fig. 5.4 Graphical illustration of the temperature controller

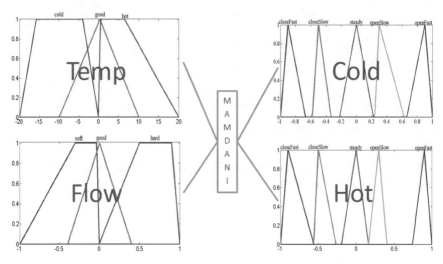

Fig. 5.5 Fuzzy system for the temperature controller

1. If (temp is cold) and (flow is soft) then (cold is openSlow)(hot is openFast)
2. If (temp is cold) and (flow is good) then (cold is closeSlow)(hot is openSlow)
3. If (temp is cold) and (flow is hard) then (cold is closeFast)(hot is closeSlow)
4. If (temp is good) and (flow is soft) then (cold is openSlow)(hot is openSlow)
5. If (temp is good) and (flow is good) then (cold is steady)(hot is steady)
6. If (temp is good) and (flow is hard) then (cold is closeSlow)(hot is closeSlow)
7. If (temp is hot) and (flow is soft) then (cold is openFast)(hot is openSlow)
8. If (temp is hot) and (flow is good) then (cold is openSlow)(hot is closeSlow)
9. If (temp is hot) and (flow is hard) then (cold is closeSlow)(hot is closeFast)

Fig. 5.6 Fuzzy rules of the temperature controller

The fuzzy system of this controller contains two inputs, the first one is called *Temperature* and is composed of three membership functions, which are called *cold, good and hot*, the two membership functions are trapezoidal and the central one is triangular. The second input is called *Flow* and is composed of three membership functions which are called *soft* is of trapezoidal type, *good* is of triangular type and *hard* trapezoidal type.

The combination of rules shown in Fig. 5.6 simulates the speed of the water flow at the desired temperature.

5.3 Mobile Robot Controller

The third control problem is the case of a mobile robot, in this case the plant is of a unicycle mobile robot [4], consisting of two driving wheels located on the same axis and a front free wheel, and Fig. 5.7 shows a graphical description of the robot model.

The robot body is symmetrical around the perpendicular axis and the center of mass is at the geometric center of the body. It has two driving wheels that are fixed to the axis that passes through the center of mass "C" represented by {C, Xm, Ym}, and one passive wheel that prevents the robot from tipping over as it moves on a plane.

The mathematical model of the robot is given by the following Eq. 5.2:

$$M(q)\dot{v} + C(q, \dot{q})v + Dv = \tau + P(t) \tag{5.2}$$

where,

$q = (x, y, \theta)^T$ is the vector of the configuration coordinates.
$v = (v, w)^T$ is the vector of velocities.
$\tau = (\tau_1, \tau_2)$ is the vector of torques applied to the wheels of the robot where τ_1 and τ_2 denote the torques of the right and left wheel, respectively.
$P \in R^2$ is the uniformly bounded disturbance vector.
$M(q) \in R^{2X2}$ is the positive-definite inertia matrix.
$C(q, \dot{q})\vartheta$ is the vector of centripetal and Coriolis forces.
$D \in R^{2X2}$ is a diagonal positive-definite damping matrix.

The kinematic system is represented by Eq. 5.3:

Fig. 5.7 Mobile robot model

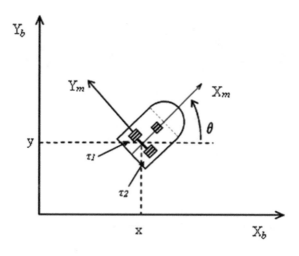

$$\dot{q} = \begin{bmatrix} \cos\theta & 0 \\ \sin\theta & 0 \\ 0 & 1 \end{bmatrix} \begin{bmatrix} v \\ w \end{bmatrix} \tag{5.3}$$

where,

(x, y) is the position in the $X - Y$ (world) reference frame

θ is the angle between the heading direction and the x-axis

v *and* w are the linear and angular velocities, respectively.

Equation 5.4 represents the non-holonomic constraint, which this system has, which corresponds to a non-slip wheel condition preventing the robot from moving sideways.

$$\dot{y}\cos\theta - \dot{x}\sin\theta = 0 \tag{5.4}$$

Figure 5.8 illustrates the fuzzy system for the robot controller, and this controller is composed of two inputs and two outputs. The first input is *ev* (error in the linear velocity), and the second is *ew* (error in the angular velocity), which have three membership functions with linguistic values of N, Z and P. The first output is t1 (torque 1) and the second output is t2 (torque 2) with three membership functions with the same linguistic values, is of Mamdani type and Fig. 5.9 shows the fuzzy rules.

The system fails to meet Brockett's necessary condition for feedback stabilization, which implies that no continuous static state-feedback controller exists that can stabilize the closed-loop system around the equilibrium point.

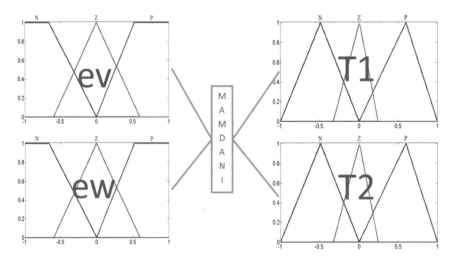

Fig. 5.8 Fuzzy system for the robot controller

```
1. If (ev is N) and (ew is N) then (T1 is N)(T2 is N)
2. If (ev is N) and (ew is Z) then (T1 is N)(T2 is Z)
3. If (ev is N) and (ew is P) then (T1 is N)(T2 is P)
4. If (ev is Z) and (ew is N) then (T1 is Z)(T2 is N)
5. If (ev is Z) and (ew is Z) then (T1 is Z)(T2 is Z)
6. If (ev is Z) and (ew is P) then (T1 is Z)(T2 is P)
7. If (ev is P) and (ew is N) then (T1 is P)(T2 is N)
8. If (ev is P) and (ew is Z) then (T1 is P)(T2 is Z)
9. If (ev is P) and (ew is P) then (T1 is P)(T2 is P)
```

Fig. 5.9 Rules of the robot controller

5.4   Inverted Pendulum Control

This control problem, unlike the aforementioned ones, it is of the Takagi-Sugeno type, which makes the level of complexity increase. In addition complexity is also originating from the nonlinear nature of the problem. The objective of this problem is to keep the pendulum in balance without falling off the car. This plant consists of a straight-line rail, a cart, a pendulum, and a driving unit. The cart can move left or right on the rail freely. The pendulum is hinged on the center of the top surface of the cart and can rotate around the pivot in the same vertical plane with the rail. Given that no friction exists in the system, Eqs. 5.5 and 5.6 represents the mathematical model of the inverted pendulum [5].

$$\propto = (m_c + m_p)g \sin \theta - \{F + m_p l_p \omega^2 \sin \theta\} \cos \theta \{4/3(m_c + m_p) - m_p(\cos \theta)^2\}l_p, \tag{5.5}$$

$$a = 4/3\{F + m_p l_p \omega^2 \sin \theta\} \\ - m_p g \sin \theta \cos \theta \{4/3(m_c + m_p) - m_p(\cos \theta)^2\} \tag{5.6}$$

The parameters m_c and m_p are, respectively, the mass of the cart and the mass of the pendulum in the unit [kg], and g $= 9.8$ m/s^2 is the gravity acceleration. The parameter l_p is the length from the center of the pendulum to the pivot in the unit [m] and equals to the half-length of the pendulum. The variable F represents the driving force in the unit [N] applied horizontally to the cart. The variables θ, ω, α represent, respectively, the angle of the pendulum from upright position, its angular velocity, its angular acceleration, and the clockwise direction is positive, respectively. The variables x, v, a denote the position of the cart from the rail origin, its velocity, its acceleration, and right direction is positive.

Figure 5.10 shows the main idea of the inverted pendulum controller.

The fuzzy system structure of the controller is shown in Fig. 5.11, which contains 4 inputs which are *Pendulum angle, Angular velocity, Cart position and Cart velocity*, all membership functions are Gaussian type, and one output that contains 16 linear functions.

The rule set for this controller is shown in Fig. 5.12, which simulates the vehicle's behavior on the cart to maintain balance.

The combination of the rules of Fig. 5.12 is given by the following meanings:

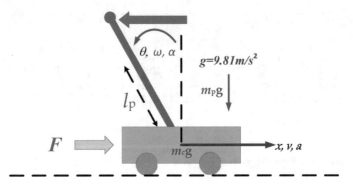

Fig. 5.10 Inverted pendulum controller

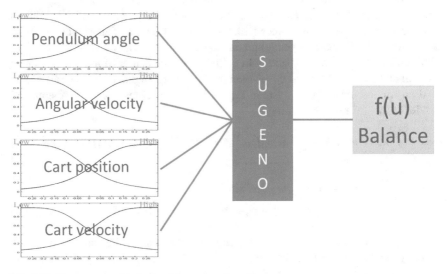

Fig. 5.11 Fuzzy system for the pendulum control problem

in1 = Pendulum angle	in1mf1 = Low
in2 = Angular velocity	in1mf2 = High
in3 = Car position	in2mf1 = Low
in4 = Car velocity	in2mf2 = High
	in3mf1 = Low
	in3mf2 = High
	in4mf1 = Low
	in4mf2 = High

They are represented the 16 Sugeno coefficients in the outputs corresponding to the combination of the rules shown in Fig. 5.12.

1. If (in1 is in1mf1) and (in2 is in2mf1) and (in3 is in3mf1) and (in4 is in4mf1) then (out is outmf1)
2. If (in1 is in1mf1) and (in2 is in2mf1) and (in3 is in3mf1) and (in4 is in4mf2) then (out is outmf2)
3. If (in1 is in1mf1) and (in2 is in2mf1) and (in3 is in3mf2) and (in4 is in4mf1) then (out is outmf3)
4. If (in1 is in1mf1) and (in2 is in2mf1) and (in3 is in3mf2) and (in4 is in4mf2) then (out is outmf4)
5. If (in1 is in1mf1) and (in2 is in2mf2) and (in3 is in3mf1) and (in4 is in4mf1) then (out is outmf5)
6. If (in1 is in1mf1) and (in2 is in2mf2) and (in3 is in3mf1) and (in4 is in4mf2) then (out is outmf6)
7. If (in1 is in1mf1) and (in2 is in2mf2) and (in3 is in3mf2) and (in4 is in4mf1) then (out is outmf7)
8. If (in1 is in1mf1) and (in2 is in2mf2) and (in3 is in3mf2) and (in4 is in4mf2) then (out is outmf8)
9. If (in1 is in1mf2) and (in2 is in2mf1) and (in3 is in3mf1) and (in4 is in4mf1) then (out is outmf9)
10. If (in1 is in1mf2) and (in2 is in2mf1) and (in3 is in3mf1) and (in4 is in4mf2) then (out is outmf10)
11. If (in1 is in1mf2) and (in2 is in2mf1) and (in3 is in3mf2) and (in4 is in4mf1) then (out is outmf11)
12. If (in1 is in1mf2) and (in2 is in2mf1) and (in3 is in3mf2) and (in4 is in4mf2) then (out is outmf12)
13. If (in1 is in1mf2) and (in2 is in2mf2) and (in3 is in3mf1) and (in4 is in4mf1) then (out is outmf13)
14. If (in1 is in1mf2) and (in2 is in2mf2) and (in3 is in3mf1) and (in4 is in4mf2) then (out is outmf14)
15. If (in1 is in1mf2) and (in2 is in2mf2) and (in3 is in3mf2) and (in4 is in4mf1) then (out is outmf15)
16. If (in1 is in1mf2) and (in2 is in2mf2) and (in3 is in3mf2) and (in4 is in4mf2) then (out is outmf16)

Fig. 5.12 Rules of the pendulum control problem

$$outmf_1 : 41.37\,\mathbf{in}1 + 10.03\,\mathbf{in}2 + 3.162\,\mathbf{in}3 + 4.288\,\mathbf{in}4 + 0.3386$$

$$outmf_2 : 40.41\,\mathbf{in}1 + 10.05\,\mathbf{in}2 + 3.162\,\mathbf{in}3 + 4.288\,\mathbf{in}4 + 0.2068$$

$$outmf_3 : 41.37\,\mathbf{in}1 + 10.03\,\mathbf{in}2 + 3.162\,\mathbf{in}3 + 4.288\,\mathbf{in}4 + 0.3386$$

$$outmf_4 : 41.41\,\mathbf{in}1 + 10.05\,\mathbf{in}2 + 3.162\,\mathbf{in}3 + 4.288\,\mathbf{in}4 + 0.2068$$

$$outmf_5 : 38.56\,\mathbf{in}1 + 10.18\,\mathbf{in}2 + 3.162\,\mathbf{in}3 + 4.288\,\mathbf{in}4 - 0.04893$$

$$outmf_6 : 37.6\,\mathbf{in}1 + 10.15\,\mathbf{in}2 + 3.162\,\mathbf{in}3 + 4.288\,\mathbf{in}4 - 0.1807$$

$$outmf_7 : 38.56\,\mathbf{in}1 + 10.18\,\mathbf{in}2 + 3.162\,\mathbf{in}3 + 4.288\,\mathbf{in}4 - 0.04893$$

$$outmf_8 : 37.6\,\mathbf{in}1 + 10.15\,\mathbf{in}2 + 3.162\,\mathbf{in}3 + 4.288\,\mathbf{in}4 - 0.1807$$

$$outmf_9 : 37.6\,\mathbf{in}1 + 10.15\,\mathbf{in}2 + 3.162\,\mathbf{in}3 + 4.288\,\mathbf{in}4 + 0.1807$$

$$outmf_{10} : 38.56\,\mathbf{in}1 + 10.18\,\mathbf{in}2 + 3.162\,\mathbf{in}3 + 4.288\,\mathbf{in}4 + 0.04891$$

$$outmf_{11} : 37.6\,\mathbf{in}1 + 10.15\,\mathbf{in}2 + 3.162\,\mathbf{in}3 + 4.288\,\mathbf{in}4 + 0.1807$$

$$outmf_{12} : 38.56\,\mathbf{in}1 + 10.18\,\mathbf{in}2 + 3.162\,\mathbf{in}3 + 4.288\,\mathbf{in}4 + 0.04892$$

$$outmf_{13} : 40.41 \, \textbf{in}1 + 10.05 \, \textbf{in}2 + 3.162 \, \textbf{in}3 + 4.288 \, \textbf{in}4 - 0.2068$$

$$outmf_{14} : 41.37 \, \textbf{in}1 + 10.03 \, \textbf{in}2 + 3.162 \, \textbf{in}3 + 4.288 \, \textbf{in}4 - 0.3386$$

$$outmf_{15} : 40.41 \, \textbf{in}1 + 10.05 \, \textbf{in}2 + 3.162 \, \textbf{in}3 + 4.288 \, \textbf{in}4 - 0.2068$$

$$outmf_{16} : 41.37 \, \textbf{in}1 + 10.03 \, \textbf{in}2 + 3.162 \, \textbf{in}3 + 4.288 \, \textbf{in}4 - 0.3386$$

5.5 Experimentation with Control Problems Using a Fuzzy System One Input and One Output

The method used to perform the experiments is as follows, the Differential Evolution algorithm initializes all parameters, and for the case where the F parameter is dynamic the fuzzy system performs that task (Type-1 or Interval Type-2 as appropriate). Subsequently the algorithm searches for the best structure for the membership functions of the controller and finally the controller is simulated with the structure proposed by the algorithm, and Fig. 5.13 illustrates the general idea of how the complete process is working.

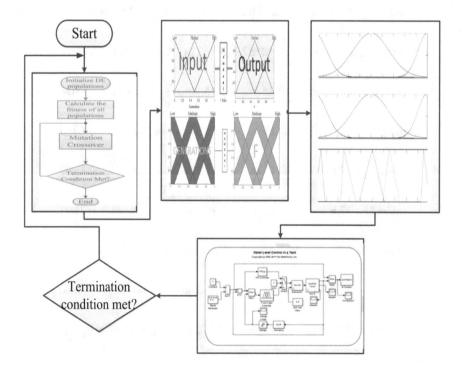

Fig. 5.13 Fuzzy controller experimentation

For the experiments we use the proposed method, which has as main dynamic adjustment the F (mutation) parameter of the original algorithm, which is recommended to be used in a range of [0, 1]. The proposed method makes the F parameter to be moved dynamically during the execution of the algorithm within the range [0, 1]. For this book we use our proposed method to change the parameter values of the membership functions for each of the fuzzy controllers. A comparison is made where we dynamically move the F parameter, first with Type-1 fuzzy logic and then with Interval Type-2 fuzzy system, which is in charge of dynamically moving the F parameter.

Figure 5.14 shows the flowchart of the original algorithm and using Type-1 and Interval Type-2 fuzzy logic.

Four fuzzy systems which are explained separately were developed, two fuzzy systems to modify the F parameter and two fuzzy systems for the CR parameter.

The Type-1 fuzzy system that we are using is defined as follows:

- Contains one input and one output
- Is of Mamdani type.
- All functions are triangular.
- The input of the fuzzy system is the generation.
- The output of the fuzzy system is the F variable.
- The fuzzy system uses 3 rules and what it does is decrease the F variable in a range of (0.1).

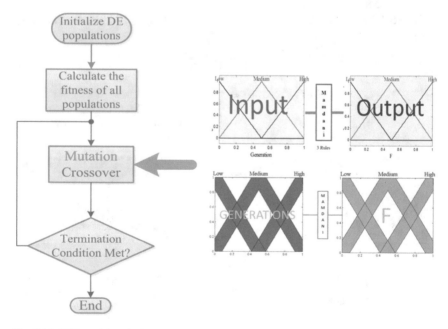

Fig. 5.14 Differential evolution with F parameter dynamic

Table 5.1 Parameters of the functions of type-1

Parameters of the membership functions for the type 1 fuzzy system	
Low	$$trimf(x,a,b,c) = \begin{cases} 0, & x \leq -0.5 \\ \frac{x+0.5}{0+0.5}, & -0.5 \leq x \leq 0 \\ \frac{0.5-x}{0.5-0}, & 0 \leq x \leq 0.5 \\ 0, & 0.5 \leq x \end{cases}$$
Medium	$$trimf(x,a,b,c) = \begin{cases} 0, & x \leq 0 \\ \frac{x-0}{0.5-0}, & 0 \leq x \leq 0.5 \\ \frac{1-x}{1-0.5}, & 0.5 \leq x \leq 1 \\ 0, & 1 \leq x \end{cases}$$
High	$$trimf(x,a,b,c) = \begin{cases} 0, & x \leq 0.5 \\ \frac{x-0.5}{1-0.5}, & 0 \leq x \leq 0.5 \\ \frac{1.5-x}{1.5-1}, & 1 \leq x \leq 1.5 \\ 0, & 1.5 \leq x \end{cases}$$

Equation 5.7 represents the mathematical form of the membership functions that form the fuzzy system and Eq. 4.2 expresses the way in which the input of the fuzzy system is calculated with which the F parameter is obtained as output, and in this way it can be used in the algorithm.

Table 5.1 contains the mathematical knowledge and Fig. 5.15 illustrates the fuzzy system.

$$trimf(x,a,b,c) = \begin{cases} 0, & x \leq a \\ \frac{x-a}{b-a}, & a \leq x \leq b \\ \frac{c-x}{c-b}, & b \leq x \leq c \\ 0, & c \leq x \end{cases} \tag{5.7}$$

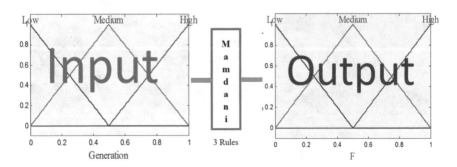

Fig. 5.15 Fuzzy system modifying dynamically F

The fuzzy rules modifying dynamically F decreased are presented in Fig. 5.16.

In the same way our Interval Type-2 fuzzy system contains one input and one output; the mathematical Eq. 5.8 represents the knowledge of the membership function and Eq. 4.2 expresses the way in which the input of the fuzzy system is calculated with which the F parameter is obtained as output. Table 5.2 contains the parameters of the functions Interval Type-2 fuzzy system.

$$\widetilde{\mu}(x) = \left[\underline{\mu}(x), \overline{\mu}(x)\right] = \text{itristype2}(x, [a_1, b_1, c_1, a_2, b_2, c_2])$$

where $a_1 < a_2, b_1 < b_2, c_1 < c_2$

$$\mu_1(x) = \max\left(\min\left(\tfrac{x-a_1}{b_1-a_1.}, \tfrac{c_1-x}{c_1-b_1}\right), 0\right)$$
$$\mu_2(x) = \max\left(\min\left(\tfrac{x-a_2}{b_2-a_2}, \tfrac{c_2-x}{c_2-b_2}\right), 0\right)$$
$$\overline{\mu}(x) = \max(\mu 1(x), \mu 2(x)) \forall x \notin (b1, b2) \tag{5.8}$$
$$\overline{\mu}(x) = 1 \forall x \in (b1, b2)$$
$$\underline{\mu}(x) = \min(\mu 1(x), \mu 2(x))$$

Fig. 5.16 Rules of the fuzzy system F decreased

| 1.- If (Generations is Low) then (F is High) |
| 2.- If (Generations is Medium) then (F is Medium) |
| 3.- If (Generations is High) then (F is Low) |

Table 5.2 Parameters of the functions of interval type-2 fuzzy system

Parameters of the membership functions for the Interval type-2 fuzzy system

Low	$\mu_1(x) = \max\left(\min\left(\tfrac{x-0.5}{-0.08+0.5}, \tfrac{0.4-x}{0.4+0.08}\right), 0\right)$
	$\mu_2(x) = \max\left(\min\left(\tfrac{x+0.4}{0.08+0.4}, \tfrac{0.5-x}{0.5-0.08}\right), 0\right)$
	$\overline{\mu}(x) = \max(\mu 1(x), \mu 2(x)) \forall x \notin (-0.08, 0.08)$
	$\overline{\mu}(x) = 1 \forall x \in (-0.08, 0.08)$
	$\underline{\mu}(x) = \min(\mu 1(x), \mu 2(x))$
Medium	$\mu_1(x) = \max\left(\min\left(\tfrac{x+0.83}{0.4+0.83}, \tfrac{0.92-x}{0.92-0.4}\right), 0\right)$
	$\mu_2(x) = \max\left(\min\left(\tfrac{x-0.08}{0.5-0.08}, \tfrac{1.07-x}{1.07-0.5}\right), 0\right)$
	$\bar{\mu}(x) = \max(\mu 1(x), \mu 2(x)) \forall x \notin (0.4, 0.5)$
	$\bar{\mu}(x) = 1 \forall x \in (0.4, 0.5)$
	$\underline{\mu}(x) = \min(\mu 1(x), \mu 2(x))$
High	$\mu_1(x) = \max\left(\min\left(\tfrac{x-0.4}{0.92-0.4}, \tfrac{1.4-x}{1.4-0.92}\right), 0\right)$
	$\mu_2(x) = \max\left(\min\left(\tfrac{x-0.5}{1.07-0.5}, \tfrac{1.5-x}{1.5-1.07}\right), 0\right)$
	$\overline{\mu}(x) = \max(\mu 1(x), \mu 2(x)) \forall x \notin (0.92, 1.07)$
	$\bar{\mu}(x) = 1 \forall x \in (0.92, 1.07)$

The Interval Type-2 fuzzy system that we are using is defined as follows:

- Contains one input and one output.
- Is of Mamdani type.
- All functions are triangular.
- The input of the fuzzy system is generation.
- The output of the fuzzy system is the variable F.
- The fuzzy system uses 3 rules and what it does is decrease the variable F in a range of (0.1).

Figure 5.17 shows the Interval Type-2 fuzzy system used to dynamically change the F variable and the rules of the fuzzy system are shown in Fig. 5.16.

For each of the control problems the objective function is defined by the calculation of the mean squared error (RMSE), which is expressed in Eq. 5.9

$$RMSE = \sqrt{\frac{1}{N} \sum_{t=1}^{N} \left(x_t - \hat{x}_t \right)^2} \tag{5.9}$$

The presented results are the average of 30 experiments performed for each of the variants. The methodology of our work is the following, we performed experiments with the original algorithm, then we used Type-1 fuzzy system to modify the F variable (mutation) and finally we used Interval Type-2 fuzzy system to modify the F variable, and to validate the efficiency of the proposed method we add noise to each of the control problems and experiments are also performed with Type-1 and Interval Type-2 fuzzy systems.

Table 5.3 shows the parameters for each of the variants of the algorithms where D is the number of dimensions, NP is the number of elements in the population, F is the mutation which is dynamic for the Type-1 and the Interval Type-2 Fuzzy systems, CR is the crossover parameter, G is the number of generations and Noise represents the level the noise in the controller plant and this noise is a uniform random number value in decibels.

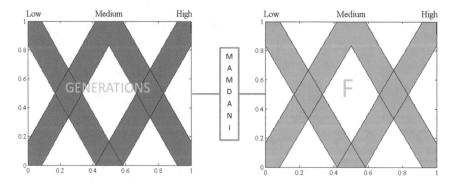

Fig. 5.17 Interval type-2 fuzzy system

Table 5.3 Parameters of the algorithm

Parameters						
	D	NP	F	CR	GEN	Noise
DE	30	50	0.5	0.5	50	–
DE + T1FS	30	50	Dynamic	0.5	50	–
DE + T1FS + N	30	50	Dynamic	0.5	50	0.50
DE + IT2FS	30	50	Dynamic	0.5	50	–
DE + IT2FS + N	30	50	Dynamic	0.5	50	0.50

In this paper we use the following abbreviations for each one of the variants with which the experiments were performed:

- DE: Differential Evolution
- DE + T1FS: Differential Evolution with Type-1 fuzzy system
- DE + T1FS + N: Differential Evolution with Type-1 fuzzy system plus noise
- DE + T2FS: Differential Evolution with Interval Type-2 fuzzy system
- DE + T2FS + N: Differential Evolution with Interval Type-2 fuzzy system plus noise.

Tables 5.4, 5.5, 5.6 and 5.7 represent the results for each of the control problems described above, and in each table the best result, the worst results, the averages and the standard deviations are presented.

Table 5.4 Results for the water tank controller

Water tank controller					
RMSE					
	DE	DE+T1FS	DE+T1FS+N	DE+IT2FS	DE+IT2FS+N
Best	1.38E−02	2.45E−04	5.66E−03	5.40E−02	1.23E−03
Worst	2.45E−01	1.21E−01	6.37E−02	6.50E−02	1.46E−01
Mean	1.12E−01	3.40E−02	6.00E−02	6.02E−02	3.22E−02
Std	5.98E−02	3.00E−02	2.46E−03	2.45E−03	3.63E−02

Table 5.5 Results for the temperature controller

Temperature controller					
RMSE					
	DE	DE+T1FS	DE+T1FS+N	DE+IT2FS	DE+ IT2FS+N
Best	5.42E−02	5.40E−02	5.40E−03	5.40E−02	5.66E−03
Worst	6.49E−02	6.49E−02	6.48E−02	6.50E−02	6.37E−02
Mean	6.11E−02	6.18E−02	6.14E−02	6.02E−02	6.00E−02
Std	2.09E−03	2.51E−03	2.68E−03	2.45E−03	2.46E−03

Table 5.6 Results for the mobile robot controller

Mobile robot controller

RMSE

	DE	DE+T1FS	DE+T1FS+N	DE+IT2FS	DE+IT2FS+N
Best	6.37E + 00	6.37E−01	1.15E−02	6.37E−03	6.37E−04
Worst	1.40E + 00	1.40E + 00	7.92E−01	1.40E + 00	1.40E + 00
Mean	2.12E + 01	2.06E−01	1.91E−01	3.62E−02	2.12E−03
Std	2.81E + 01	2.01E−01	1.73E−01	3.18E−02	2.91E−03

Table 5.7 Results for the inverted pendulum control

Inverted Pendulum control

RMSE

	DE	DE+T1FS	DE+T1FS+N	DE+IT2FS	DE+IT2FS+N
Best	5.84E−01	1.02E−01	4.33E−02	1.83E−02	1.40E−02
Worst	2.46E + 00	1.40E + 00	1.26E + 00	1.40E + 00	4.33E−01
Mean	1.49E + 00	4.30E−01	3.30E−01	2.15E−01	2.41E−01
Std	5.20E−01	3.62E−01	2.69E−01	2.75E−01	1.65E−01

Figure 5.18 shows the plots for each the best results obtained for the water tank controller, where the blue line represents the trajectory to follow and the pink line is the result obtained using the algorithm with each of its variants.

Figure 5.19 shows the behavior of the best result of the different experiments performed with the water tank controller. We can observe that the Interval Type-2 fuzzy systems without noise and with noise are better than the original DE algorithm and the algorithm using Type-1 fuzzy system.

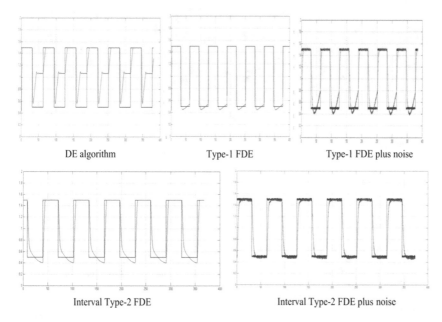

| DE algorithm | Type-1 FDE | Type-1 FDE plus noise |

| Interval Type-2 FDE | Interval Type-2 FDE plus noise |

Fig. 5.18 Simulation of results for the water tank

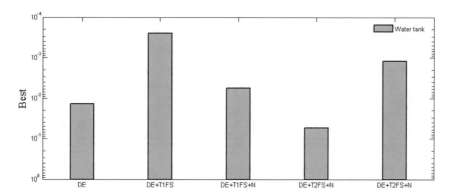

Fig. 5.19 Plot of the best result

Figure 5.20 shows the plots for each of the best results obtained for the temperature controller, where the blue line represents the trajectory to follow and the pink line is the result obtained using the algorithm with each of its variants.

For the case of the temperature controller it is observed that Interval Type-2 fuzzy system with noise is better, in comparison to the other variations, although the difference between all the results is minimal and the plot in Fig. 5.21 illustrates all the best results.

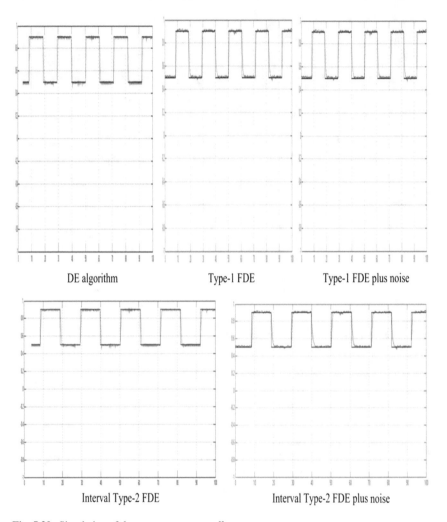

Fig. 5.20 Simulation of the temperature controller

Table 5.6 represents the results obtained for the experiments carried out with the mobile robot controller, and Figs. 5.22 and 5.23 illustrate the simulation of the best results obtained for each of the variations made, then a plot of the best results obtained by each one of the variations is presented, respectively.

Table 5.7 represents the results obtained for the experiments carried out with the inverted pendulum controller, and Figs. 5.24 and 5.25 illustrate the simulation of the best results obtained for each of the variations made, then a plot of the best results obtained by each one of the variations is presented, respectively.

We can clearly notice in both plots that the results obtained for this controller are good since the optimization of the algorithm using fuzzy logic is better than the

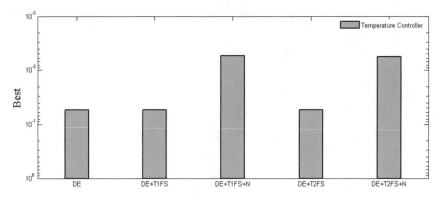

Fig. 5.21 Plot of the best results

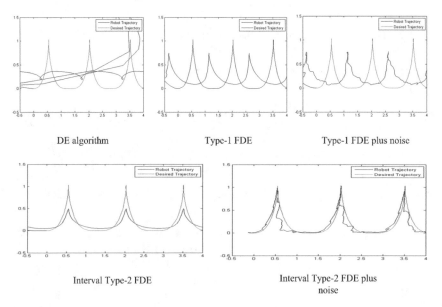

Fig. 5.22 Simulation results for the mobile robot controller

original algorithm. Figure 5.26 illustrates the mean for each of the controllers, and we can observe the performance of the original algorithm and the variants proposed for the algorithm with Type-1 and Interval Type-2 fuzzy systems.

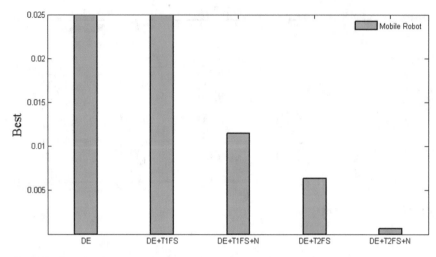

Fig. 5.23 Plot of the best error for the mobile robot controller

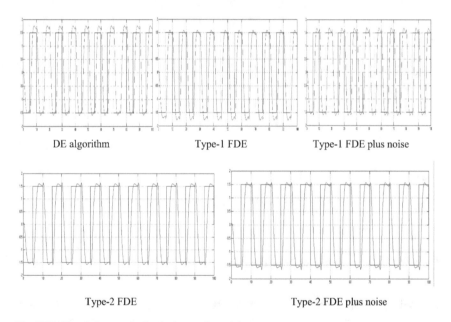

Fig. 5.24 Simulation results for the inverted pendulum

5.6   Statistical Tests with Control Problems

To verify the results of both methods and to provide a statement of which method is better, a statistical test was performed. The Z statistical test of two samples was used to make a comparison between the original Differential Evolution algorithm

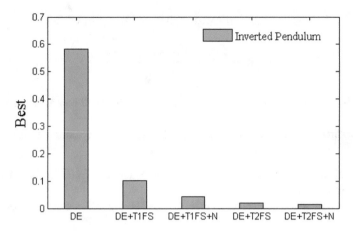

Fig. 5.25 Plot of the best error for the inverted pendulum control

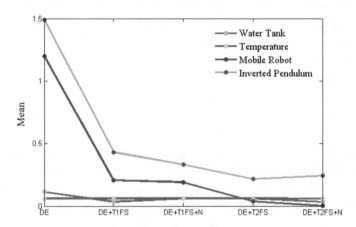

Fig. 5.26 Plot for the means for the four controllers

and the Fuzzy Differential Evolution with Type-1 and Interval Type-2 fuzzy systems. The statistical test used for comparison is the z-test, whose parameters are defined in Table 5.8.

Table 5.8 Parameters for the statistical testing

Parameter	Value
Level of significance	95%
Alpha	5%
H_0	$\mu_1 \geq \mu_2$
H_a	$\mu_1 < \mu_2$ (claim)
Critical value	-1.645

The Eq. 5.10 is for the test that was applied and is expressed follows:

$$Z = \frac{(\overline{X_1} - \overline{X_2}) - (\mu_1 - \mu_2)}{\sigma_{\overline{X_1} - \overline{X_2}}} \tag{5.10}$$

In this case, 12 statistical tests were made, three for each controller, and a brief explanation of these tests is presented below:

In the first statistical test we compare the Differential Evolution algorithm with the Fuzzy Differential Evolution algorithm where:

- The null hypothesis states that the average of the DE+T1FS is greater than or equal to the average of the DE.
- The alternative hypothesis states that the DE+1FS algorithm average is lower than the average of the DE.

In the second statistical test we compare the Type-1 Fuzzy Differential Evolution algorithm with the Interval Type-2 Fuzzy Differential Evolution algorithm where:

- The null hypothesis states that the average of the DE+T2FS is greater than or equal to the average of the DE+T1FS.
- The alternative hypothesis states that the DE+T2FS algorithm average is lower than the average of the DE +T1FS.

In the third statistical test we compare the Fuzzy Differential Evolution algorithm plus noise with Interval Type-2 Fuzzy Differential Evolution algorithm plus noise where:

- The null hypothesis states that the average of the DE+ T2FS+N is greater than or equal to the average of the DE +T1FS+N.
- The alternative hypothesis states that the DE +T2FS + N algorithm average is lower than the average of the DE+T1FS+N.

The region of rejection is defined for all values below—1.645.

The Eq. 5.10 expresses the formula used, the data from the values of the mean and standard deviation for the original method and the proposed method are obtained from Tables 5.4, 5.5, 5.6 and 5.7, respectively.

Table 5.9 contains the results for each of the Z values and indicates whether the statistical test is significant or not for each controller.

Table 5.9 clearly shows that the obtained results are favorable compared to the original algorithm, except for two of the nine tests performed, where we did not obtain a favorable result, but in general, it can be can concluded that the use of Interval Type-2 fuzzy logic is better in all four controllers used.

To complement our study, a statistical test is carried out with other methods and thus to know the competitiveness we have compared with other algorithms that combine fuzzy logic to make dynamic some or several of the parameters. For the case of the water tank, temperature and mobile robot controllers we make a statistical comparison with the harmony search algorithm [6].

Table 5.9 Statistical tests for the four case studies

Statistical tests				
Case study	μ_1	μ_2	z value	Evidence
Water tank controller	DE + T1FS	DE	−6.3857	Significant
	DE + T2FS	DE + IT1FS	4.7676	Not Significant
	DE + T2FS + N	DE + IT1FS + N	−4.1851	Significant
Temperature controller	DE + T1FS	DE	−0.5031	Not significant
	DE + T2FS	DE + IT1FS	−2.4985	Significant
	DE + T2FS + N	DE + IT1FS + N	−2.1079	Significant
Mobile robot controller	DE + T1FS	DE	−4.092	Significant
	DE + T2FS	DE + IT1FS	−4.5702	Significant
	DE + T2FS + N	DE + IT1FS + N	−5.9791	Significant
Inverted pendulum controller	DE + T1FS	DE	−9.1633	Significant
	DE + T2FS	DE + IT1FS	−2.5904	Significant
	DE + T2FS + N	DE + IT1FS + N	−1.5447	Not significant

Tables 5.10, 5.11 and 5.12 show the statistical tests between both methods using an Interval Type-2 fuzzy system to make dynamic a parameter of each algorithm, the parameters for the statistical tests are those contained in Table 5.8.

In the case of inverted pendulum controller, the comparison is made with another reference but with the same harmony search algorithm [8], statistical tests are performed using the Type-1 fuzzy system for comparison and the parameters of the statistical test used are contained in Table 5.7. Table 5.13 shows the statistical test for the inverted pendulum controller.

Table 5.10 Statistical tests for the water tank controller case study

Water tank controller				
Method	Mean	Standard deviation	z-value	Evidence
DE + IT2FS	6.02E−02	2.45E−03	9.5464	Not significant
FHS2 [7]	2.56E−02	1.97E−02		
DE + IT2FS + N	3.22E−02	3.63E−02	2.6392	Not significant
FHS2 + N [7]	1.32E−02	1.254E−02		

Table 5.11 Statistical tests for the Temperature Controller case study

Temperature controller				
Method	Mean	Standard deviation	z-Value	Evidence
DE + IT2FS	6.02E−02	2.45E−03	−4.6692	Significant
FHS2 [7]	6.25E−02	1.13E−03		
DE + IT2FS + N	6.00E−02	2.46E−03	132.0655	Not significant
FHS2 + N [7]	6.29E−04	1.07E−04		

Table 5.12 Statistical tests for the mobile robot case study

Mobile robot controller

Method	Mean	Standard deviation	z-Value	Evidence
DE + IT2FS	3.62E−02	3.18E−02	−2.9130	Significant
FHS2 [7]	1.11E−01	1.37E−01		
DE + IT2FS + N	2.12E−03	2.91E−03	−4.5248	Significant
FHS2 + N [7]	3.69E−02	4.20E−02		

Table 5.13 Statistical tests for the inverted pendulum controller case study

Inverted pendulum controller

Method	Mean	Standard deviation	z-Value	Evidence
DE + T1FS	4.30E−01	3.62E−01	−1.2107	Significant
FHS [8]	7.67E−01	4.81E−01		
DE + T1FS + N	3.30E−01	2.69E−01	−4.4369	Significant
FHS + N [8]	7.54E−01	4.49E−01		

Two different types of statistical tests were performed, the first where when the controller has no noise and the second where the controller has noise for each of the plants, and of these statistical tests we can conclude that for the more complex problems, that in this case are the mobile robot controller and inverted pendulum controller our methodology obtains better results.

5.7   D.C. Motor Speed Controller

The speed control in a D.C. Motor is a classical benchmark control problem with real applications in the industry, and this problem is proposed in this paper as a study case because consists on the optimization of a real problem. So, this experiment aims to demonstrate the capability of the proposed approach to improve the optimization of the controller, and in this case, the experiment is focused on reducing the optimization time. Figure 5.27 shows the graphical representation of speed control in a D.C. Motor [9].

The fuzzy system for the speed control in a D.C. Motor contains two inputs and one output, input 1 is the error and input 2 is the change of the error, and the output is the voltage. The system is of Mamdani type, as shown in Fig. 5.28. The fuzzy system contains 15 fuzzy rules, which are shown in Table 5.14.

The reference for the motor speed is illustrated in Fig. 5.29, where the control objective is to move from a resting state to a reference speed of 40 m/s.

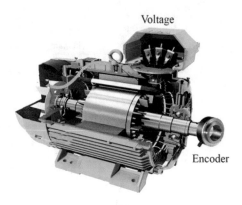

Fig. 5.27 Graphical representation

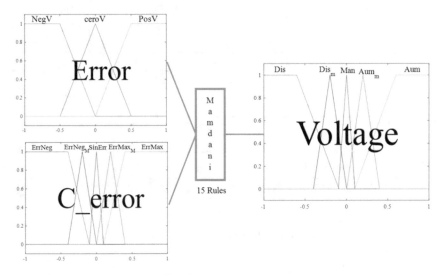

Fig. 5.28 Fuzzy system for controlling the motor

The DE + FS algorithms were applied for the optimization of the fuzzy system for speed control of the motor, using the same parameters, population, generations, dimensions, number of experiments F and CR, which are described in Table 5.15.

The objective function of the methods is the root mean square error (RMSE) of the reference for the motor, as shown in Eq. 5.9.

This section describes the structures of fuzzy systems, the inputs and outputs, rules and types of membership functions that are used to combine with the Differential Evolution algorithm.

The fuzzy systems used are composed of two inputs and two outputs, all are Mamdani type, have 9 rules and then describe each of them:

Table 5.14 Fuzzy rules for the controller

No.	Inputs		Output
	Error	Change in error	Voltage
1	NegV	ErrNeg	Dis
2	NegV	SinErr	Dis
3	NegV	ErrMax	Dis_m
4	CeroV	ErrNeg	Aum_m
5	CeroV	ErrMax	Dis_m
6	PosV	ErrNeg	Aum_m
7	PosV	SinErr	Aum
8	PosV	ErrMax	Aum
9	CeroV	SinErr	Man
10	NegV	ErrNeg_M	Dis
11	CeroV	ErrNeg_M	Aum_m
12	PosV	ErrNeg_M	Aum
13	PosV	ErrMax_M	Aum
14	CeroV	ErrMax_M	Dis_m
15	NegV	ErrMax_M	Dis

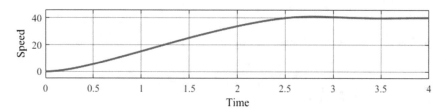

Fig. 5.29 Speed response without optimization

Table 5.15 Parameters of DE + FS algorithm

Parameter	DE + FS
Population	50
Dimensions	45
Generations	30
Number of experiments	30
F	Dynamic
Cr	Dynamic

- Type-1 fuzzy logic system

The Type 1 fuzzy system is illustrated in the Fig. 5.30 which indicates that two input and two output are composed of triangular membership functions type and

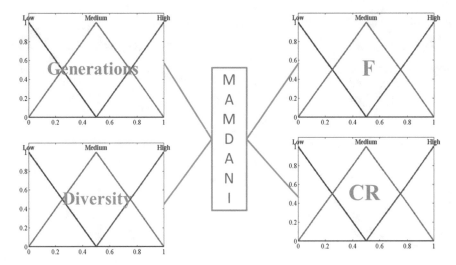

Fig. 5.30 Type-1 fuzzy logic system

are granulated in low, medium and high, the mathematical equation of membership functions is described in Eq. 5.7.

The parameters corresponding to the mathematical equation are expressed in the Table 5.16 which summarizes and represents the parameters of the inputs and outputs of the Type 1 fuzzy system.

Table 5.16 Parameters of the type 1 fuzzy logic sets

Type 1 fuzzy logic sets	
Low	$trimf(x,a,b,c) = \begin{cases} 0, & x \le -0.5 \\ \frac{x+0.5}{0+0.5}, & -0.5 \le x \le 0 \\ \frac{0.5-x}{0.5-0}, & 0 \le x \le 0.5 \\ 0, & 0.5 \le x \end{cases}$
Medium	$trimf(x,a,b,c) = \begin{cases} 0, & x \le 0 \\ \frac{x-0}{0.5-0}, & 0 \le x \le 0.5 \\ \frac{1-x}{1-0.5}, & 0.5 \le x \le 1 \\ 0, & 1 \le x \end{cases}$
High	$trimf(x,a,b,c) = \begin{cases} 0, & x \le 0.5 \\ \frac{x-0.5}{1-0.5}, & 0 \le x \le 0.5 \\ \frac{1.5-x}{1.5-1}, & 1 \le x \le 1.5 \\ 0, & 1.5 \le x \end{cases}$

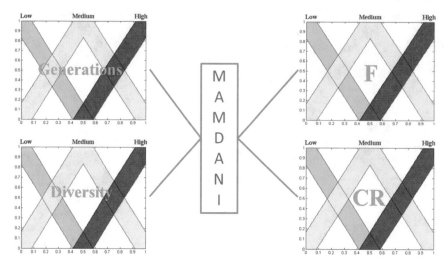

Fig. 5.31 Interval type 2 fuzzy logic system

- Interval Type-2 fuzzy logic system

The Interval Type 2 fuzzy system is illustrated in the Fig. 5.31 which indicates that two inputs and two outputs are composed of triangular membership functions type and are granulated in low, medium and high, the mathematical equation of membership functions is described in Eq. 5.8.

In the case of this fuzzy system it is important that the FOU for each of the membership functions is symmetric as can be seen in Fig. 5.31. The parameters corresponding to the mathematical equation are expressed in the Table 5.17 which summarizes and represents the parameters of the inputs and outputs of the Type 1fuzzy system

- Generalized Type 2 fuzzy systems

The Generalized Type-2 fuzzy logic system is illustrated in the Fig. 5.32 which indicates that two input and two output are composed of triangular membership functions type and are granulated in low, medium and high, the mathematical equation of membership functions is described in Eq. 5.11.

Table 5.17 Table parameters of the interval type 2 fuzzy logic sets

Interval type-2 fuzzy logic sets	
Low	$\mu_1(x) = \max\left(\min\left(\frac{x-0.5}{-0.08+0.5}, \frac{0.4-x}{0.4+0.08}\right), 0\right)$
	$\mu_2(x) = \max\left(\min\left(\frac{x+0.4}{0.08+0.4}, \frac{0.5-x}{0.5-0.08}\right), 0\right)$
	$\overline{\mu}(x) = \max(\mu_1(x), \mu_2(x)) \forall_x \notin (-0.08, 0.08)$
	$\overline{\mu}(x) = 1 \ \forall_x \in (-0.08, 0.08)$
	$\underline{\mu}(x) = \min(\mu_1(x), \mu_2(x))$
Medium	$\mu_1(x) = \max\left(\min\left(\frac{x+0.83}{0.4+0.83}, \frac{0.92-x}{0.92-0.4}\right), 0\right)$
	$\mu_2(x) = \max\left(\min\left(\frac{x-0.08}{0.5-0.08}, \frac{1.07-x}{1.07-0.5}\right), 0\right)$
	$\overline{\mu}(x) = \max(\mu_1(x), \mu_2(x)) \forall_x \notin (0.4, 0.5)$
	$\overline{\mu}(x) = 1 \ \forall_x \in (0.4, 0.5)$
	$\underline{\mu}(x) = \min(\mu_1(x), \mu_2(x))$
High	$\mu_1(x) = \max\left(\min\left(\frac{x-0.4}{0.92-0.4}, \frac{1.4-x}{1.4-0.92}\right), 0\right)$
	$\mu_2(x) = \max\left(\min\left(\frac{x-0.5}{1.07-0.5}, \frac{1.5-x}{1.5-1.07}\right), 0\right)$
	$\overline{\mu}(x) = \max(\mu_1(x), \mu_2(x)) \forall_x \notin (0.92, 1.07)$
	$\overline{\mu}(x) = 1 \ \forall_x \in (0.92, 1.07)$
	$\underline{\mu}(x) = \min(\mu_1(x), \mu_2(x))$

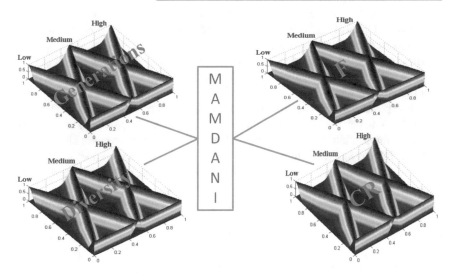

Fig. 5.32 Generalized type-2 fuzzy logic system

$$\mu(x, u) = trigausstype2(x, u[a_1, b_1, c_1, a_2, b_2, c_2, \rho])$$

$$\mu(x, u) = exp\left[-\frac{1}{2}\left(\frac{u-P_x}{\sigma_u}\right)\right] where$$

$$\mu_1(x) = \max\left(\min\left(\frac{x-a_1}{b_1-a_1}, \frac{c_1-x}{c_1-b_1}\right), 0\right) and$$

$$\mu_2(x) = \max\left(\min\left(\frac{x-a_2}{b_2-a_2}, \frac{c_2-x}{c_2-b_2}\right), 0\right)$$

$$\overline{\mu}(x) = \begin{cases} \max(\mu_1(x), \mu_2(x)) \ \forall_x \notin (b_1, b_2) \\ 1 \qquad\qquad\qquad\quad \forall_x \in (b_1, b_2) \end{cases} \tag{5.11}$$

$$\underline{\mu}(x) = \min(\mu_1(x), \mu_2(x))$$

$$\rho_x = \max\left(\min\left(\frac{x-a_x}{b_x-a_x}\right), \left(\frac{c_x-x}{c_x-b_x}\right), 0\right), where$$

$$a_x = \frac{a_1+a_2}{2}, b_x = \frac{b_1+b_2}{2}, c_x = \frac{c_1+c_2}{2},$$

$$\delta = \overline{\mu}(x) - \underline{\mu}(x)$$

$$\sigma_u = \frac{1+\rho}{2\sqrt{3}}\delta + \varepsilon$$

where a_1, b_1 and c_1 parameters are for the upper membership function and a_2, b_2 and c_2 are for the lower membership function, respectively, where ρ is fraction of uncertainty of the support for the secondary membership function.

The mathematical knowledge of the fuzzy system is described in the Table 5.18 which is summarized and represents the parameters of the inputs and outputs of the fuzzy system.

The results are presented in a comparison where the controller is used without noise (FLC without noise) and subsequently applying noise in the controller (FLC with noise) of control applying noise of a uniform random number generator for each of the different fuzzy systems used.

The comparison presented in Table 5.19 corresponds to the simulation of control problem with the original algorithm and the problem of control applying noise using the original algorithm.

The experimentation obtained with the original algorithm with respect to the controller is better but when applying noise in the controller we can see that the method cannot stabilize the controller voltage and good results are not obtained.

Table 5.20 present the results using the Type 1 fuzzy system, without noise in the FCL and with noise in the FCL, the table contains the best result, the worst result, the mean and standardized deviation.

When using the Type-1 fuzzy system, can we observed with respect to the original algorithm a significant improvement, in the same way it is observed that noise is added in the controller the result obtained is better.

Table 5.21 present the results using the Interval Type 2 fuzzy system, without noise in the FCL and with noise in the FCL, the table contains the best result, the worst result, the mean and standardized deviation.

Table 5.22 present the results using the Generalized Type-2 fuzzy logic system, without noise in the FCL and with noise in the FCL, the table contains the best result, the worst result, the mean and standardized deviation.

To verify the results of both methods and to provide a statement of which method is better, a statistical test was performed. The Z statistical test of two samples was

Table 5.18 Table parameters of the generalized type-2 fuzzy logic sets

Generalized type-2 fuzzy logic sets	
Low	$\mu_1(x) = \max\left(\min\left(\frac{x-0.5}{-0.08+0.5}, \frac{0.4-x}{0.4+0.08}\right), 0\right)$ and $\mu_2(x) = \max\left(\min\left(\frac{x+0.4}{0.08+0.4}, \frac{0.5-x}{0.5-0.08}\right), 0\right)$ $\overline{\mu}(x) = \begin{cases} \max(\mu_1(x), \mu_2(x)) \ \forall_x \notin (-0.08, 0.08) \\ 1 \qquad\qquad\qquad \forall_x \in (-0.08, 0.08) \end{cases}$ $\underline{\mu}(x) = \min(\mu_1(x), \mu_2(x))$ $\rho_x = \max\left(\min\left(\frac{x-a_x}{b_x-a_x}\right), \left(\frac{c_x-x}{c_x-b_x}\right), 0\right)$, where $a_x = \frac{-0.5-0.4}{2}, b_x = \frac{-0.8-0.08}{2}, c_x = \frac{-0.4-0.5}{2}$, $\delta = \overline{\mu}(x) - \underline{\mu}(x)$ $\sigma_u = \frac{1+\rho}{2\sqrt{3}}\delta + \varepsilon$ where $\rho = 0.5$
Medium	$\mu_1(x) = \max\left(\min\left(\frac{x+0.084}{0.4+0.084}, \frac{0.92-x}{0.92-0.4}\right), 0\right)$ and $\mu_2(x) = \max\left(\min\left(\frac{x-0.084}{0.5-0.084}, \frac{1.07-x}{1.07-0.5}\right), 0\right)$ $\overline{\mu}(x) = \begin{cases} \max(\mu_1(x), \mu_2(x)) \ \forall_x \notin (0.4, 0.5) \\ 1 \qquad\qquad\qquad \forall_x \in (0.4, 0.5) \end{cases}$ $\underline{\mu}(x) = \min(\mu_1(x), \mu_2(x))$ $\rho_x = \max\left(\min\left(\frac{x-a_x}{b_x-a_x}\right), \left(\frac{c_x-x}{c_x-b_x}\right), 0\right)$, where $a_x = \frac{-0.084+0.084}{2}, b_x = \frac{0.4-0.5}{2}, c_x = \frac{0.92-1.09}{2}$, $\delta = \overline{\mu}(x) - \underline{\mu}(x)$ $\sigma_u = \frac{1+\rho}{2\sqrt{3}}\delta + \varepsilon$ where $\rho = 0.5$
High	$\mu_1(x) = \max\left(\min\left(\frac{x-0.4}{0.92-0.4}, \frac{1.4-x}{1.4-0.92}\right), 0\right)$ and $\mu_2(x) = \max\left(\min\left(\frac{x-0.5}{1.07-0.5}, \frac{1.5-x}{1.5-1.07}\right), 0\right)$ $\overline{\mu}(x) = \begin{cases} \max(\mu_1(x), \mu_2(x)) \ \forall_x \notin (0.92, 1.07) \\ 1 \qquad\qquad\qquad \forall_x \in (0.92, 1.07) \end{cases}$ $\underline{\mu}(x) = \min(\mu_1(x), \mu_2(x))$ $\rho_x = \max\left(\min\left(\frac{x-a_x}{b_x-a_x}\right), \left(\frac{c_x-x}{c_x-b_x}\right), 0\right)$, where $a_x = \frac{0.4+0.5}{2}, b_x = \frac{0.92-1.07}{2}, c_x = \frac{1.4-1.5}{2}$, $\delta = \overline{\mu}(x) - \underline{\mu}(x)$ $\sigma_u = \frac{1+\rho}{2\sqrt{3}}\delta + \varepsilon$ where $\rho = 0.5$

Table 5.19 Original algorithm FLC with and without noise

DE FLC without noise		DE FLC with noise	
Best	6.33E−01	Best	9.81E−01
Worst	7.34E−01	Worst	9.96E−01
Mean	6.99E−01	Mean	9.91E−01
Standard deviation	4.78E−02	Standard deviation	6.61E−03

Table 5.20 FDE algorithm FLC with and without noise

FDE FLC without noise		FDE FLC with noise	
Best	9.53E−01	Best	6.47E−01
Worst	9.71E−01	Worst	6.73E−01
Mean	9.60E−01	Mean	6.64E−01
Standard deviation	5.33E−03	Standard deviation	1.18E−02

Table 5.21 IT2FDE algorithm FLC with and without noise

IT2FDE FLC without noise		IT2FDE FLC with noise	
Best	4.55E−01	Best	9.78E−02
Worst	6.35E−01	Worst	9.92E−01
Mean	4.95E−01	Mean	6.29E−01
Standard deviation	5.39E−02	Standard deviation	4.78E−02

Table 5.22 GT2FDE algorithm FLC with and without noise

GT2FDE FLC without noise		GT2FDE FLC with noise	
Best	4.47E−02	Best	5.33E−03
Worst	5.14E−01	Worst	5.33E−01
Mean	2.80E−01	Mean	1.21E−01
Standard deviation	2.05E−01	Standard deviation	1.68E−01

used to make a comparison between the original Differential Evolution algorithm and the Fuzzy Differential Evolution with Type-1 and interval type-2 fuzzy logic.

The statistical test used for comparison is the z-test, whose parameters are defined in Table 5.8.

In this case 4 statistical tests were made for the controller, and a brief explanation of these tests is presented below:

In the first statistical test we compare the Differential Evolution algorithm FLC without noise with the Differential Evolution algorithm FLC with noise where:

- The null hypothesis states that the average of the DER is greater than or equal to the average of the DE.

- The alternative hypothesis states that the DER algorithm average is lower than the average of the DE.

In the second statistical test we compare the Type-1 Fuzzy Differential Evolution algorithm FLC without noise with the Type-1 Fuzzy Differential Evolution FLC with noise where:

- The null hypothesis states that the average of the FDER is greater than or equal to the average of the FDE.
- The alternative hypothesis states that the FDER algorithm average is lower than the average of the FDE.

In the third statistical test we compare the Fuzzy Differential Evolution algorithm FLC without noise with Interval Type-2 Fuzzy Differential Evolution algorithm FLC without noise:

- The null hypothesis states that the average of the T2FDER is greater than or equal to the average of the T2FDE.
- The alternative hypothesis states that the T2FDER algorithm average is lower than the average of the T2FDE.

In the fourth statistical test we compare the Generalized Type-2 Fuzzy Differential Evolution algorithm FLC without noise with Generalized Type-2 Fuzzy Differential Evolution algorithm FLC without noise:

- The null hypothesis states that the average of the GT2FDER is greater than or equal to the average of the GT2FDE.
- The alternative hypothesis states that the GT2FDER algorithm average is lower than the average of the GT2FDE.

In all cases the region of rejection is for all values below—1.645.

Table 5.23 contains the results for each the Z values and indicates whether the statistical test is significant or not for each controller

Table 5.23 clearly shows that the obtained results are favorable compared to the original algorithm, except for one of the four tests performed, where we did obtain a favorable result, but in general, it can be can concluded that the use of the fuzzy logic system is better in controller used.

Table 5.23 Statistical tests for the case studies

Statistical tests				
Case study	μ_1	μ_2	z value	Evidence
Speed control of the motor	DER	DE	33.1438	Not significant
	FDER	FDE	−125.2137	Significant
	IT2FDER	IT2FDE	−4.0356	Significant
	GT2FDER	GT2FDE	−3.2858	Significant

References

1. Fierro R, Castillo O (2013) Design of fuzzy control systems with different PSO variants. In: Castillo O, Melin P, Kacprzyk J (eds) Recent advances on hybrid intelligent systems, vol 451. Springer Berlin Heidelberg, Berlin, Heidelberg, pp 81–88
2. Amador-Angulo L, Castillo O (2017) Comparative analysis of designing differents types of membership functions using bee colony optimization in the stabilization of fuzzy controllers. In: Melin P, Castillo O, Kacprzyk J (eds) Nature-inspired design of hybrid intelligent systems, vol 667. Springer International Publishing, Cham, pp 551–571
3. Castillo O, Amador-Angulo L, Castro JR, Garcia-Valdez M (2016) A comparative study of type-1 fuzzy logic systems, interval type-2 fuzzy logic systems and generalized type-2 fuzzy logic systems in control problems. Inf Sci 354:257–274
4. Sanchez MA, Castillo O, Castro JR (2015) Generalized type-2 fuzzy systems for controlling a mobile robot and a performance comparison with interval type-2 and type-1 fuzzy systems. Expert Syst Appl 42(14):5904–5914
5. Caraveo C, Valdez F, Castillo O (2016) Optimization of fuzzy controller design using a new bee colony algorithm with fuzzy dynamic parameter adaptation. Appl Soft Comput 43:131–142
6. Peraza C, Valdez F, Melin P (2017) Optimization of intelligent controllers using a type-1 and interval type-2 fuzzy harmony search algorithm. Algorithms 10(3):82
7. Castillo O, Ochoa P, Soria J (2016) Differential evolution with fuzzy logic for dynamic adaptation of parameters in mathematical function optimization. In: Angelov P, Sotirov S (eds) Imprecision and uncertainty in information representation and processing, vol 332. Springer International Publishing, 2016, pp 361–374
8. Castillo O, Valdez F, Soria J, Amador-Angulo L, Ochoa P, Peraza C (2019) Comparative study in fuzzy controller optimization using Bee colony, differential evolution, and harmony search algorithms. Algorithms 12(1):9
9. Ontiveros E, Melin P, Castillo O (2018) Impact study of the footprint of uncertainty in control applications based on interval type-2 fuzzy logic controllers. In: Fuzzy logic augmentation of neural and optimization algorithms: theoretical aspects and real applications. Springer, Cham, pp 181–197

Chapter 6
Conclusions

The augmentation of DE performance by using fuzzy logic is the main contribution of this book. To this end a set of experiments was performed by using Type-1, Interval Type-2 and Generalized Type-2 fuzzy systems to be able to make a comparative study of the DE performance. In this case the different Differential Evolution variants were applied to a set of Benchmark control problems, and as a conclusion of our work we can state that the use of fuzzy logic to optimize the DE parameters improves performance, and for this book we have shown this with statistical tests.

The first conclusion is that the use of Type-1 or Interval Type-2 fuzzy systems in DE provides better results than the original algorithm, although it will depend a lot on the particular problem and the measure of performance, since as we can note for the case of the water tank controller the use of a Type-1 fuzzy system was better than the Interval Type-2 fuzzy system.

In addition, drawn from the study is that for problems with a higher level of uncertainty the use of Interval Type-2 fuzzy system is better since in this work in the four case studies when we considered noise for the controller the use of Interval Type-2 fuzzy system obtained significant evidence of better results with respect to a Type-1 fuzzy system.

In a general way we can conclude that the proposal to use fuzzy logic combined with the differential evolution algorithm is a good one, since statistically it is verified that the use of an Interval Type-2 fuzzy system to dynamically change some parameter is better than the original algorithm. In the same way statistically, it is verified that when compared with other fuzzy algorithms it obtains better results.

In this book a comparison of results between the different fuzzy systems was performed in order to determine the behavior of the Differential Evolution algorithm. In particular, there is not previous use in the literature of the Differential Evolution algorithm and a Generalized Type-2 fuzzy logic applied to control problems. We can statistically state that the use of the GT2FDE algorithm is better than the use of the original algorithm and in turn that the proposed T1FDE and IT2FDE algorithms and this was statistically shown. As for the means of all the experiments it can be

O. Castillo et al., *Differential Evolution Algorithm with Type-2 Fuzzy Logic for Dynamic Parameter Adaptation with Application to Intelligent Control*, SpringerBriefs in Computational Intelligence, https://doi.org/10.1007/978-3-030-62133-9_6

observed that they are of the same order, the difference is noticeable when we observe the best results for each of the cases with and without noise in the controller.

It can be noted in the experiments how the results are improving when using each of the different fuzzy systems, the application of noise in the controllers shows us that the improvement is remarkable, and this confirms that the use of fuzzy logic with greater uncertainty can improve the results. It is important to mention that more experimentation and more control cases are needed by applying the proposed methodology to be able to confirm that for a group or several types of controllers our methodology obtains good results, we can say that we have a good favorable start to our proposed methodology.

Appendix

For the all case studies was used the Matlab programming language for the code for the Differential evolution Optimization Algorithm applied to Fuzzy Controllers. In this book, the code for the simulation that shows the solution in the representation for the second studies case (Autonomous Robot Mobile Controller) is shown below.

Main Differential Evolution Algorithm for Type-1 FLC and Interval Type-2 FLC applied to the D.C. Motor

© The Author(s), under exclusive license to Springer Nature Switzerland AG 2021   53
O. Castillo et al., *Differential Evolution Algorithm with Type-2 Fuzzy Logic for Dynamic Parameter Adaptation with Application to Intelligent Control*,
SpringerBriefs in Computational Intelligence,
https://doi.org/10.1007/978-3-030-62133-9

```
%% BENCHMARK D.C. Motor Speed Controller %%%%
clear all
close all
clc

%Read the original fis
Motor = readfis('Motor');
save Motor;

%    Read the diffuse to move the parameters

%fis=readfis('FDEtriangulares');%Type-1 fuzzy logic
%it2fis= readfistype2('FDETrianType2');% Interval Type-2 fuzzy logic
fis=readgfistype2('FDEg1.fis'); % Generalized Type-2 fuzzy logic
  tic;

u=0.8; % Noise level
D1 = 45; % dimension of problem
NP = 45; % size of population
% F = 0.8; % differentiation constant
% CR = 0.3; % crossover constant
GEN = 25; % number of generations

objfun='CreaFisMotor'; %Objetive function

D=45; %Number of points of membership functions

fprintf('Star simulation');
runtime=30;% The algorithm can bbe excuted repeatedly

erroresbco=[];
Errores = zeros(1,8);
Mutacion = zeros(runtime);   %%**Mutation
Cruce = zeros(runtime);      %%**Crossover
mejormsee = zeros(runtime);
```

```
% ** ALGORITHM'S VARIABLES ** %
X = zeros(1,D1); % trial vector
%Pop = zeros(D1,NP); % population
Fit = zeros(1,NP); % fitness of the population
 iBest = 1; % index of the best solution
r = zeros(3,1); % randomly selected indices

Pop=zeros(NP,D1);

for a=1:runtime

    %%PARA EL TANK
        ub=ones(1,D)*1;     %upper limit
      lb=ones(1,D)*-(1);  % lower limit
    tic;
%/*las varibles son inicializadas en el rango de [lb,ub]. si cada
parametro es de rango diferente, use arrays lb[j], ub[j]
Range = repmat((ub-lb),[NP 1]);
Lower = repmat(lb, [NP 1]);
individual = rand(NP,D) .* Range + Lower;

for i=1:NP
%
    for d=1:D1
%
        Pop(i,d)=unifrnd(lb(d),ub(d));
      end
end

  [par,pos]=min(Pop);%%%%%%
  D2(a)=diversidad(Pop,pos);%%%%

for cf=1:D1
      [Motor individual] =feval(objfun,individual,cf);
%       sim('PlantaMotor'); % Call the plant without noise
        sim('PlantaMotorRuidoP'); % Call the plant without noise
         rmsee= sqrt(sum((Error(:)).^2)/numel(Error));
        ObjVal(cf,:)= rmsee;

end;

Fit=ObjVal;
BestInd=find(ObjVal(:)==min(ObjVal(:)));
BestInd = min(BestInd(:));
GlobalMin=ObjVal(BestInd);
GlobalParams=individual(BestInd,:);
Gen=1;
while ((Gen <= GEN)),
i=1;
t=1;
  while(t<=NP)% for t = 1:NP

  %*************************Type-1 Fuzzu Logic****************************
%   nor=Gen/GEN;
%     F=evalfis(nor,fis);
```

```
      F=out(1);
%        CR=out(2);

  %*************************Interval Type-2 Fuzzy logic******************
%    nor=Gen/GEN;
%       out=evalfis(nor,fis);
%       F=out(1);
%       CR=out(2);

  %********************** Generalized Type-2 Fuzzy Logic **************
Diver(i)=D2(i);
 nor=Gen/GEN;
       out=evalgfistype2([nor;D2(i)],fis);
        F=out(1);
       CR=out(2);

%*****************************************************************************

          r(1) = floor(rand()* NP) + 1;
          while r(1)==t
          r(1) = floor(rand()* NP) + 1;
          end
          r(2) = floor(rand()* NP) + 1;
          while (r(2)==r(1))||(r(2)==t)
          r(2) = floor(rand()* NP) + 1;
          end
          r(3) = floor(rand()* NP) + 1;
          while (r(3)==r(2))||(r(3)==r(1))||(r(3)==t)
          r(3) = floor(rand()* NP) + 1;
          end
      Rnd = floor(rand()*D1) + 1;
          for ii = 1:D1

              if ( rand()<CR ) || ( Rnd==ii )
                  %fprintf('Error111 =%g\n',r(1),r(2),r(3));
                  X(ii) = Pop(ii,r(3)) + F * (Pop(ii,r(1)) - Pop(ii,r(2)));
                  %fprintf('Error111 =%g\n',F);

              else
                  %
                  X(ii) = Pop(ii,t);
              end
          end

        sol=(ub-lb).*rand(1,D)+lb;
        sol=individual(i,:);

        cont=1;

        Motor=feval(objfun,sol,cont);
%        sim('PlantaMotor'); %)
         sim('PlantaMotorRuidoP');

rmsee= sqrt(sum((Error(:)).^2)/numel(Error));
        ObjValSol(t)= rmsee;
```

```
            FitnessSol=ObjValSol;

        for x=1:(D1)
          if (FitnessSol(t)<Fit(x))
                individual(x,:)=sol;
                Fit(x)=FitnessSol(t);
                ObjVal(x)=ObjValSol(t);
                X(t)=0;
            else
                X(i)=X(i)+1; %/* If the solution cannot be improved, increase
your test counter
            end;
        end;
      t=t+1;
  end;
i=i+1;


            ind=find(ObjVal(:)==min(ObjVal(:)));
            ind = min(ind(:));
            if (ObjVal(ind)<GlobalMin)
            GlobalMin=ObjVal(ind);
            GlobalParams=individual(ind,:);
            baf= individual(ind,:); %% Save the best fis1 architecture
            bestFis{a,1}= Motor; %% Save the best fis architecture for the
D.C Motor

            mejormsee(Gen) = rmsee;
            end;
            fprintf('Generaciones=%d Numero de corridas  %d Valor
Objetivo=%g\n',Gen,a,GlobalMin);
            disp(strcat('  Best = ',num2str(Fit(iBest)),' Time =
',num2str(toc), strcat('F = ',num2str(F), ' CR = ',num2str(CR) )));
        D2(Gen+1)= diversidad(Pop,pos);
            Gen=Gen+1;
  end;

GlobalMins(a)=GlobalMin;
tiempo(a)= toc;
minimoError = min(GlobalMins(:));
promError = mean(GlobalMins(:));
maximo = max(GlobalMins(:));
mejormse= min(mejormsee(:));
wps{a,1} = Motor;
tiempo(a)= toc/60;
Errores(a,:) = [itaev itsev iaev isev minimoError maximo promError
mejormse];

end;% end of the algorithm

tiempo2=tiempo';
ERRORED = GlobalMins';
save wps;
save GlobalMins;
```

```
%Fuzzy System Controller
[System]
Name='Motor1'
Type='mamdani'
Version=2.0
NumInputs=2
NumOutputs=1
NumRules=15
AndMethod='min'
OrMethod='max'
ImpMethod='min'
AggMethod='max'
DefuzzMethod='centroid'

[Input1]
Name='Error'
Range=[-1 1]
NumMFs=3
MF1='NegV':'trapmf',[-2 -1 -0.7976 -0.3259]
MF2='ceroV':'trimf',[-0.786 0 0.688]
MF3='PosV':'trapmf',[-0.3208 0.7074 1 2]
[Input2]
Name='C_Error'
Range=[-1 1]
NumMFs=5
MF1='ErrNeg':'trapmf',[-4 -0.8621 -0.6297 -0.5466]
MF2='SinErr':'trimf',[-0.5258 0 0.4025]
MF3='ErrMax':'trapmf',[0.4187 0.6628 0.7974 4]
MF4='ErrNeg_M':'trimf',[-0.6274 -0.5822 0]
MF5='ErrMax_M':'trimf',[0 0.5311 0.6462]
[Output1]
Name='Voltaje'
Range=[-1 1]
NumMFs=5
MF1='Dis':'trapmf',[-1 -0.9151 -0.8266 -0.3309]
MF2='Man':'trimf',[-0.3759 0.1972 0.5059]
MF3='Aum':'trapmf',[0.3293 0.7333 0.8276 1]
MF4='Dis_m':'trimf',[-0.6092 -0.5501 0.1567]
MF5='Aum_m':'trimf',[0.2294 0.5741 0.6739]
[Rules]
1 1, 1 (1) : 1
1 2, 1 (1) : 1
1 3, 4 (1) : 1
2 1, 5 (1) : 1
2 3, 4 (1) : 1
3 1, 5 (1) : 1
3 2, 3 (1) : 1
3 3, 3 (1) : 1
2 2, 2 (1) : 1
1 4, 1 (1) : 1
2 4, 5 (1) : 1
3 4, 3 (1) : 1
3 5, 3 (1) : 1
2 5, 4 (1) : 1
1 5, 1 (1) : 1
```

```
% Interval Type-2 Fuzzy System to dynamically move the parameters
[System]
Name='DoSalidas'
Type='mamdani'
Version=2.0
NumInputs=2
NumOutputs=2
NumRules=9
AndMethod='min'
OrMethod='max'
ImpMethod='min'
AggMethod='max'
DefuzzMethod='centroid'
[Input1]
Name='Generations'
Range=[0 1]
NumMFs=3
MF1='Bajo':'itritype2',[-0.5833 -0.08333 0.4167 -0.4167 0.08333 0.5833]
MF2='Medio':'itritype2',[-0.08353 0.4157 0.9149 0.08347 0.5827 1.082]
MF3='Alto':'itritype2',[0.4167 0.9167 1.417 0.5833 1.083 1.583]
[Input2]
Name='Diversity'
Range=[0 1]
NumMFs=3
MF1='Bajo':'itritype2',[-0.5833 -0.08333 0.4167 -0.4167 0.08333 0.5833]
MF2='Medio':'itritype2',[-0.08333 0.4167 0.9167 0.08333 0.5833 1.083]
MF3='Alto':'itritype2',[0.4167 0.9167 1.417 0.5833 1.083 1.583]

[Output1]
Name='F'
Range=[0 1]
NumMFs=3
MF1='Bajo':'itritype2',[-0.5833 -0.08333 0.4167 -0.4167 0.08333 0.5833]
MF2='Medio':'itritype2',[-0.08353 0.4157 0.9149 0.08347 0.5827 1.082]
MF3='Alto':'itritype2',[0.4167 0.9167 1.417 0.5833 1.083 1.583]

[Output2]
Name='Cr'
Range=[0 1]
NumMFs=3
MF1='Bajo':'itritype2',[-0.5833 -0.08333 0.4167 -0.4167 0.08333 0.5833]
MF2='Medio':'itritype2',[-0.08333 0.4167 0.9167 0.08333 0.5833 1.083]
MF3='Alto':'itritype2',[0.4167 0.9167 1.417 0.5833 1.083 1.583]

[Rules]
1 1, 3 1 (1) : 1
2 1, 2 2 (1) : 1
3 1, 1 3 (1) : 1
1 2, 3 1 (1) : 1
2 2, 2 2 (1) : 1
3 2, 1 3 (1) : 1
1 3, 3 1 (1) : 1
2 3, 2 2 (1) : 1
3 3, 1 3 (1) : 1
```

Index

© The Author(s), under exclusive license to Springer Nature Switzerland AG 2021
O. Castillo et al., *Differential Evolution Algorithm with Type-2 Fuzzy Logic for Dynamic Parameter Adaptation with Application to Intelligent Control*,
SpringerBriefs in Computational Intelligence,
https://doi.org/10.1007/978-3-030-62133-9

Printed in the United States
By Bookmasters